Reconstruction by Way of the Soil

RECONSTRUCTION
BY WAY OF THE SOIL

Guy Theodore Wrench

A DISTANT MIRROR

RECONSTRUCTION BY WAY OF THE SOIL
by Guy Wrench

Originally published in 1946.

ISBN 9780648859413

A DISTANT MIRROR
web adistantmirror.com.au
email admin@adistantmirror.com.au

Contents

Acknowledgements

IN THE CONSTRUCTION of this book, I am indebted to many living authors, whose words I have quoted in my text.

Amongst them, I feel I owe a special tribute to Mrs Elspeth Huxley for the use I have made of her book *The Red Strangers*. I have immeasurable admiration for her exquisite story.

In my faith in the primary value of the soil I have been greatly strengthened by the books of two honoured friends, the Earl of Portsmouth's *Alternative to Death* and Lord Northbourne's *Look to the Land*.

The Earl's book, published in 1943, has only recently reached me. With a general outlook closely similar to my own, the author has something which I do not possess; namely, an intimate, personal knowledge of all that pertains to the soil of Britain. His book constitutes the comprehensive guide for which all workers determined to build a sane, earth-based foundation for our national life have been looking.

Lord Northbourne's book was published in 1940, and it has been my frequent companion in the three years which I have taken in the writing of this book. Lord Northbourne has also helped me in ways surpassing the usual kindness of friendship. He has taken full charge of the typescript of the book in England, and, by a careful study of the text, has assisted me with most valuable criticisms.

Lastly, I wish to thank my friend, Dr Haji Kassim, for his great help in the compilation of the 21st and 22nd chapters.

~ *Dr Guy Theodore Wrench*

1

Introduction

IT WILL BE CLEAR to a reader, who, like a prospector confronting a face of rock, runs his eye down the page of contents of this book, that its subject is a general one. It is, indeed, widespread both in space and time, yet in spite of its generality it cannot be said to be widely recognized; so little so, in fact, that to many readers it will appear a new subject.

People, under the advanced differentiation of the present day, are apt to think of themselves as finished products – as soldiers, merchants, sailors, engineers, lawyers and so on – but to speculate on what they, one and all, actually *are*, seldom occupies many of them for more than a few casual moments.

Nevertheless, now that they are involved in a supreme crisis (*the author is referring to WW2 – ed.*), now that, however complete victory may be, the future cannot be the replica of the past, it is inconceivable that humanity will not be forced to face fundamental questions such as in previous times of habit and routine they were able to avoid.

They have already come to learn that this age, so distinguished for its scientific progress and its widespread knowledge has, in spite of these advantages, *completely failed in its promise of peace and prosperity.*

Even in such vital social problems as feeding and employment, it has failed, and failed signally. Those who have now been forced to experience in their own lives, and therefore to reflect upon, these two problems, are astounded that their resolution has been so definitely brought about by war. Where peace failed, it seems that war has succeeded brilliantly.

The will of the people and the skill of organization have assured all of their share of the national food supply. Those who do hard manual labour can rely on receiving sufficient to allow them to accomplish their work without the weariness that results from partial starvation.

Why is war so much more effective in these respects than peace? What is lacking in times of peace that comes into being in times of war?

Is it that under the supreme strain of war against a powerful and ruthless enemy there arises in the homeland a peace, goodwill, and sense of brotherhood which displaces the greedy competition, the covert hostility and the social barriers of peacetime that destroy the best qualities of country, blood and language? Does our civilization need war to make decency of human conduct prevail?

Many answers have been given to these and kindred questions, but in order to look at them afresh, it is proposed in this book to review conditions, both historical and immediate, with a vision untarnished by the pride of the present, pride attached to that in which one's ego has its being. This is a broad statement, for all are tarred with the same brush, and no one can claim impartiality exempting him from his heritage and the prejudice of circumstance. Yet, if we are to enjoy a better communal and individual life after the war than before it, the attempt has to be made with the probity it demands.

To introduce the attempt is the object of this opening chapter, and to make this beginning we will try to look at people not as final products, not as labourers, merchants, shopmen and the rest, not as rich and poor, sick and healthy, wise and foolish, but as they are – all and each inseparably linked together in a common likeness, one which will pervade the chapters of this book.

This common likeness is the fact that we are all supporting our lives with the products of the soil. Like other forms of life, vegetable and animal, we humans are dependent for our existence upon the crust of the earth on which we live.

Humanity, however, possess a marked peculiarity which distinguishes us from other forms of earthly life. It is this – that we alone have been able to make ourselves *partners* in the creative power of the soil. We alone are agriculturists or farmers, whereby we assure ourselves a constant supply of food, clothing and other primal necessities instead of having to trust in the uncertainties of chance.

Mankind alone has acquired a degree of mastery over the earth. In this ability to take part in the creation of their necessities, humanity has gained something more than a mere increase in its food supply. We have gained an understanding, dim though it may be, of the relationship between ourselves and the powers which rule the universe and that minute part of it on which we live. Humans have realized that to be partners in creation, they have to submit themselves to the unavoidable autocracy of these powers; they have to be, in their own language, creatures of the Creator, and as such, however headstrong and dominant they may be over weaker forms of life than theirs, they are, nevertheless, like them limited by the laws of their existence.

Upon the basis of limitation, humans are inevitably compelled to shape their individual and social lives. Should they transgress, they or their descendants are inevitably punished. These rules and restrictions under which mankind lives are those of the very nature of life and death.

Life and death are the two essential conditions of earthly existence; they are the two different phases of this existence.

The living may cease to be alive, but it is not lost to the cycle of existence, and remains within it as a necessary part of it.

In the condition which is called 'dead', matter is either in the soil or will eventually reach it. That which, by its life, has had the power to lift itself from the crust of the earth now returns to that crust. There it plays an essential part in promoting further life. In a word, there is no actual death as a permanent thing. There is only a suspension of life. Death itself is but a phase of life in which dead matter returns to the soil, where it is reformed into living matter again.

There is nothing that has once taken life from the soil, that will not, by reaching the soil, again become living. The dead leaf that we see lying on the path at our feet is not dead in the sense of being finished. Let it lie, and, through the creative agency of the soil, its substance will again enter into a blade of grass, a flower, an insect, bird or animal, and so return to the kingdom of the living.

Life and death are therefore not separate entities, but phases of each other. The living has to respect the dead as a part of itself, not ultimately dead but living. This respect has been expressed in the religious life of humanity by various forms of reverence in which the innate eternity of life in its most highly developed form, that of the human soul, is recognized.

When humans do not interfere and the soil is left to itself, it does not fail. Through it everything that has passed from a state of life is restored again to a state of life; nothing is lost.

In the philosophy of modern science, however, the seeds that lie scattered upon the ground and do not fructify are stigmatized as failures, while those that grow into plants are dubbed the fittest, because they survive and expand into plants. Yet the other seeds survive no less; they re-enter cycles of life by other paths. Some even enter the very plants to which their fellow seeds have given rise. So, for example, every one of the countless seeds of the elm that litter the ground in early summer and which fail to turn into trees are not failures in the symphony of nature. In a musical symphony, each note, even the lowest and lowliest, *fits*. It is not a question of the fittest excluding or making superfluous the remainder. That is a wholly false outlook upon the processes of earthly life. Each has its place without which the whole is incomplete. Each has its place in a creative cycle, each passes from soil to plant and then, in many cases, to animal, and, after an interlude of death, returns to the creative realm of soil.

This is the symphony of nature and creation to which humans as creatures of the earth are inevitably bound and yet not wholly bound. Though they themselves are products of the soil, yet through the possession of their intellect, they have become *co-creators* and, within their limited human sphere, fashioned in the image of the Creator.

They can produce life other than their own. To do this in accord with the processes of creation, they must themselves be continuous and limited in production; they must act in harmony with the process, as it exists on earth apart from them.

Here they have to *fit*; they have to act within a process of balance. In it the living as a whole are balanced by the dead as a whole. In the living itself, its chief forms, vegetable and animal, balance each other. They are interdependent, and are incomplete without each other.

In the exchange of vegetable and animal life with the enveloping atmosphere, a similar balance is effected. It has to be regarded as a whole of balanced parts and therefore is, in human phraseology, of the character of art. Nothing in it is isolated, everything belongs to the pattern. From this art of fitting within the whole, certain consequences necessarily follow.

Wholeness or health – two words of a like origin and meaning – are one consequence.

*

This wholeness as a consequence has to be proved. Though it seems logical enough, yet little has been done to prove it in an age of unprecedented speed and discovery, an age of immense progress at a constantly expanding periphery, which by distance has inured man to the earth itself.

We are today no longer whole or healthy, physically or mentally. In the careful work of the Peckham investigators, it has been established that the vast majority of us are subnormal. We have broken away from the great primary fact of our existence – that we are first and foremost *earthly animals*, and, until we regain that fact and put it into practice, we cannot expect our social and individual lives to be whole.

Our civilization, threatened with destruction as we know it to be, has to be *healed* – another word meaning *whole* – and to be healed it has to be overhauled and *reconstructed in its relation to the soil that provides it with the means of existence.*

This was the task the bare outline of which presented itself to the author when, as a medical student, I was appalled by the crowded state of the outpatient department of a large London hospital.

'*Why disease? What then is health?*' were the questions that often vexed me. To answer them I had neither the opportunity, nor the tenacity which truly great men have in pursuing an object that is to them a consuming passion and for which they will forgo the pleasures of life, and which end so often in destitution and despair.

For that heroic life I lacked the courage, but the questions did not entirely leave me. It was only when I had leisure in which to retire for a space of years (brought to an end by the war) that I was able to gather material for the answer to this fundamental question regarding correct earthly being:

'Is there a relation of humanity to the soil which assures our health?'

The answer came as a decided yes, and in the instances I was able to gather, I found that humans could acquire health if they gave to the soil on which they lived all the food and water it required, and if they did not weaken it by exceeding the limits of the creative powers which nature had allotted to it.

My chief lesson I gained from a little-known people called the Hunza, to whom I was attracted by what Sir Robert McCarrison, who knew them well, wrote of them:

> 'They are long lived, vigorous in youth and age, capable of great endurance and enjoy a remarkable freedom from disease in general.'

He found that the Hunza, isolated in their mountain valley amidst the vast mountains of the Karakoram in north-west India, gave close attention to the soil, which strangely enough, seemingly related them to a golden age of agriculture. As a strengthening of this supposition, he found that their present farming recalled to that most cultured of mountaineers, the late Lord Conway, the unsurpassed farming of pre-Spanish Peru, the remnants of which he had seen, and which caused another explorer, O. F. Cook of the Bureau of Plant Industry of the USA

Department of Agriculture, to state:

> 'Agriculture is not a lost art, but must be reckoned as one of those which reached a remarkable development in the remote past and afterwards declined.'

The glowing pages of Prescott's second chapter in the *Conquest of Peru* seem to shine again with the Hunza, huddled between the highest congress of great mountains on earth.

Cook found that the Hunza meticulously preserved the rule of return. They were, indeed, the source of my understanding of the ultimate nature of the soil and man, and of the relationship between life and death, to which I referred a few pages back. Nothing that once got life from the Hunzas' soil was ever wasted; everything, from the least fleck of wool, the fallen leaf, the broken nutshell, to human refuse itself, was gathered and after suitable preparation returned to the soil for its food.

The Hunzas paid the same heed to water, which, by means of their principal aqueduct, the Berber – itself famous in its own right – they brought along with its silt from a glacier snout to their terraced fields. Of the Berber, Lord Conway wrote:

> 'The Alps contain no *Wasserleitung* which for volume and boldness of position can be compared to the Hunza canal. It is a wonderful work for such a toolless people as the Hunza to have accomplished, and it must have been done many centuries ago and maintained ever since, for it is the life-blood of the valley.'

Here, too, they were like the people of Peru, of whose waterways, stretching for hundreds of miles across the slopes and precipices of mountains, Prescott wrote:

> 'That they should have accomplished these difficult works with such tools as they possessed is truly wonderful.'

The words 'many centuries ago' led me to further inquiries. I found that Professor N. I. Vavilov, of the Institute of Applied Botany, Leningrad, had discovered that the area of which the Hunza Valley forms a part

> 'is one of the most important and primary world agricultural centres, where the diversity of a whole series of plants have originated'.

The people of ancient Peru, according to Cook, also produced a wonderful series of plants in the secluded valleys of the Andes and so made them the most important originating agricultural centre in America.

Here, then, within the precincts of British-supervised India, was a people who brought quite a marvellous message from the remote past, a past that justifies the tradition of the Golden Age – a past of perfect relations between humanity and the soil.

The Hunzas had created a symphony of nature. As each note, however humble, has its proper place in a symphony of Beethoven, so even the humblest fallen leaf and each drop of water have their place in the symphony of Hunza. I learned from the Hunza that their work too, was art in its original sense – derived from from *aro*, to 'fit'.

I learnt that farming is an art, and something infinitely more than just scientific agriculture. *It is a way of life itself.*

So much for the health and constantly cheerful wholeness which the Hunza enjoy. There are many other examples of this health still extant on the globe, all of them in places remote from our Western civilization. To those who are interested in this – at present – novel meaning of genuine health, I commends my book *The Wheel of Health*, in which these examples are recorded in detail. It is an essential subject to understand for any who feel the need for a reconstruction by way of the soil.

Nevertheless, it must be said that such small and remote examples are scarcely likely to have much effect on those upon whom this reconstruction by way of the soil is now urged. It seems that one is destined to stir one's readers with the negative proof of the devastation and sickness that the modern era has brought to the soil and its products, rather than by isolated proofs of wholeness, health, cheerfulness and wellbeing.

Before, however, entering upon the long path of negative proof, there presents itself a second positive element of construction, which is complementary to the meticulous care of the soil. This is the *form* in which that meticulous care of the soil is undertaken. The form is that of *family farming.*

Family farming

The family as a group is but a human complement of the soil itself, both family and soil recreating life. The family is human continuity, and the soil is vital continuity. Continuity of the family necessitates marriage as the mode of the bond of the woman to the soil; marriage bringing sons and daughters to the service of the land. It is the land that gave its particular meaning to the farming family; it is its creative power that united itself with the creation of the farmers' children. Marriage, the bearing of children, the apprenticeship of children, the respect of children for their parents and their ancestors, the care that is bestowed by the elders on the present generation because it is to repeat itself in future generations – all this wholeness of life finds its true significance in continuous family ownership or inherited right to the land.

It is, then, the land as family property, or in lesser and

more dependent degree, craft as family property, requiring the work of the family for their continuity, which primarily gives stability to men and women, making a *people*.

This right the people of ancient Peru possessed. Their self-governing communities or *ayullus*, settled in ownership of limited areas of land, existed from remote antiquity. They were the basis of the autocratic state, and they themselves constituted an agrarian communism collectively holding the land. The uniting of the *ayullus* was effected by the rulership of the principal *ayullu*, or royal family.

By far the majority, too, of the Hunza families – and the Hunza are also an ancient people – are freeholders, subject in their unity to the rulership of the *Mir*.

The greatest example of family farming is to be found among the Chinese. Their empire is by far the most stable and continuous in the world's history, and it was originally founded in the long distant past upon family property, or right to the land.

It was to their revered sages that the Chinese have always attributed their *Tsing Tien* system, the system of the nine fields. A square of land was divided by drawing two lines across it from side to side and two up and down, as in the nursery game of noughts and crosses. Nine squares were thereby formed, eight outer and one central square. The eight outer squares of land were allotted to eight families, while the centre square was worked co-operatively and its produce given to the government officials as a tax in kind.

This division into nine squares was symbolic of the principles of the sages. Where it could be, it was no doubt carried out, but it was not rigid. The soil is not so uniform in character that it can be divided with such exactness. One square might be less readily cultivated than another; one family might be larger than another. So adjustments were made; for example, if one family had several sons and another none, one or more sons of the first might be adopted by the second family.

Adaptations were made, but the principal and standard measurements remained. It was considered by the sages as the principle of choice because it promoted co-operation, close social relations, mutual production, easy exchange of commodities, unified customs, saving of individual expenses, and it connected the work and life of the families to the nation as a whole through the work which the combined families undertook on the central field. This central field could also be adjusted within limits; it could be enlarged or diminished according to the general fortune of the province or nation.

The nine squares within a square symbolized a simple approach to life, which without doubt produced a stability now inconceivable to our Western minds trained in its opposites – in change, progress and instability. We have become accustomed to regarding stability as stagnation. However, since we have become confused and disillusioned with progress and the disasters which it has brought and with which it further threatens almost all mankind, we have come to think of traditional methods with more interest and approval, but nevertheless as something still distant and foreign to us.

Yet if nature is limited, and humanity cannot pass certain boundaries or exceed certain controls without bringing upon itself generations of disaster or even human extinction, then some such stable system as that of the Chinese takes upon itself a very different aspect in the measure of human wisdom.

It may be that it will then appear as a natural human system, in scale and endurance the greatest achievement in the partnership of intelligent man and nature upon the earth. It was such a system that long ago attained a certain finality, a completion such as a great art work, a great cathedral or temple reaches. The building needs care, love and daily attendance and sometimes renovation, but it cannot be made more beautiful. It reaches its excellence and, though time may make it more revered and loved, its very excellence shows that it had, from the very beginning, the power of duration within it.

All great art has this duration. It is not subject to frequent change as is science. Changes fail to improve it. Recasting a symphony of Beethoven would not make it more beautiful, but less. It is the devotion with which it is played that bestows its beauty as human generations pass.

It is in this sense that we should, I believe, try to appreciate and understand the *Tsing Tien* system. It is a national thing on a great scale *that has kept within the limits imposed by nature*. Through this system, the Chinese have produced and maintained a productivity from the soil unexcelled elsewhere, and have supported a community of peasant-family farmers, the largest in numbers, the most skilful, the most contented and the most peaceful amongst the peoples of mankind.

The Chinese have, of course, had their misfortunes and occasional catastrophes. They have been beset by people without any settled system such as they enjoyed. Large landowners have from within sometimes destroyed the rights of the peasants, but the *Tsing Tien* system has been the thread upon which has been strung period after period of their long history.

Dr Ping-Hua Lee, in Volume 99 of *Studies in History* (Columbia University), writes:

> 'The whole history of government administration of agriculture in China coincides with the history of the *Tsing Tien* system, for it started

with this system of land tenure. Its vicissitudes, its crises and epochs were timed by the abolition or re-establishment of the system ... It is fortunate for the economic historian that the history of the *Tsing Tien* system is coincident with China's political history.'

Thus in the small body of the Hunza and in the large body of the Chinese, though much disrupted by the recent and present havoc, we have rare survivals, instances of skilled and continuous life within the limits that are set by nature and the land; a skillful fitting of mankind into the life cycle of the planet.

The Chinese had not the stupendous secluding mountain wall of the Hunza, but for as far as their power could reach, they built such a wall – the Great Wall, stretching for 1,500 miles – to shut out the Tartar. They had not the control of their water supplies from their sources as had the Peruvians and the Hunza; the floods of their great rivers have their origins in huge ranges of stripped hills mostly outside their control.

Yet in spite of these foes of stability, their system endured until it was finally worn down by the constant attrition of contact with the West. Although it has been the West and its ways that have broken up this system, nevertheless sufficient of it is known, thanks to the Chinese predilection for written history, to see in it a supreme example of the Wisdom of the East in contrast to the Science of the West.

The *Tsing Tien* has been a system of a human partnership with the soil. In it was secured for century after century the comprehensive range of both the minuteness and grandeur of this partnership, which has by no Western writer been better expressed than by Hasbach in his unique *History of the English Agricultural Labourer*:

> 'Trifles are the very objects of the small cultivator; he has everything near him and under his eye, makes use of every small advantage, cultivates every corner, has the help of his wife, and brings up his children to be the most useful the country produces. Such men serve the land as it should be served, never stinting themselves, and as absorbed in their service as any priest in his religion.'

Upon this foundation stable civilizations have been built, and can be built again.

2

Rome

ROME AND ITS CIVILIZATION were the progenitors of the civilized Western world; consequently, without knowledge of Rome's relation to the soil, it would not be possible for us to extract from history the principles of reconstruction from the soil.

We must study history, because in no other way can we tell what the Roman land was like and how it looked. History reveals that if, by some magic, we could transport ourselves back to the days of the early Latin farmers, we should see a picture of a well-populated countryside, with the land divided into a large number of small farms, often as small as five acres in size.

As each small farm had to support a family, the farming was intensive, so that the fields would have appeared crowded with a variety of crops. The bulk of the family's food and also that of its domestic animals came from the farm. We would see the various members of the family hard at work upon the farm – busy upon the land itself, and in the home and dairy.

We would see also a large number of prosperous villages. Other things would also be there – things of great importance, and soon to be described – some of which would be visible to the eye, and some not.

Now let us look at the same land some five centuries later. The picture is now quite different. We should see but few villages and few small farms, and upon the farms we should see what farmers call 'foul' fields and even land that was derelict. In place of the multitude of small farms we would see mainly orchards, vineyards and dairy farms. It would, indeed, be quite clear to us that the main object of this different form of farming was to supply fruit, grapes, olives, milk and cheese to people who did not work upon the farms or in the villages at all, but who lived in Rome, the proud city that was soon to become the capital of the Mediterranean world.

We would also see that these estates were no longer worked by Italian farmers, but by quite a different sort of men – clearly not Italians, and men lacking the buoyancy and freedom of the older farmers. We should, indeed, have reason to rub our eyes, for some of these men, incredible though it might seem to us, would be shackled with iron and even chained to each other while they worked. These were slaves.

These are the two pictures we have seen; the first is that of family farming, the second that of capitalist farming.

Transporting ourselves to an even later date, we see a third picture. The land is now swampy and derelict, and its most significant products are now swarms of mosquitoes which cause the fevers that permit only a few wretched men and cattle to scrape together some sort of livelihood and that visit, with lethal effect, the inhabitants of the great but waning city of Rome itself. This last is the picture of debased soil.

Now let us see how history explains these three pictures.

*

Of the farming of their ancestors the later Romans had no history. Nevertheless, a strong tradition existed, and that tradition placed both farming and farmers very high.

In the words of the elder Cato, to call a man a good farmer was in the past the best commendation, the highest praise.

Now this praise in the pages of *De Agricultura* must have been read a host of times without more than a general significance or regretful sentiment being attached to it. But modern discovery has shown that it had a very sound, practical significance.

The high esteem of the men of ancient Italy for good farming and the facts concomitant with it were not sentimental; they have been summed up with these words:

> 'It is impossible, after surveying such elaborate undertakings, to avoid the conclusion that Latium in the sixth century B.C. was cultivated with an intensity that has seldom been equalled anywhere.'

This is the statement of a modern authority, Professor Frank. In short, the understanding of the later Romans of the wonderful farming of their ancestors was not founded upon sentiment, but upon fact. By the time of Cato and later writers, a good deal of sentiment had entered, and a good deal of fact had slipped away.

These later Romans knew that their ancestors had been great farmers, but they do not seem to have known the greatest part of their work. That has been revealed by modern investigators and particularly by the excavations of M. la Blanchère, published in 1893 in *Mémoires présentes par divers savants à l'Académie des Inscriptions et Belles Lettres.*

Professor Tenny Frank, the above-quoted authority in *An Economic History of Rome* (1927), summarizes la Blanchère's remarkable paper, which can be itself read in the library of the British Museum. The excavations reveal that Italy was the home of a farming culture which has seldom been equalled anywhere. It had a standard of farming equal to the great farming of ancient Peru, the farming of Asia Minor in its prolific days, the farming which Professor Vavilov researches, the farming of the

Hunza – the farming, indeed, of many or even all great countries of the world in a time when farming reached a great height, only to later fall fell so steeply as to become oblivious to it.

Professor Frank begins his book on the economic story of Rome where it should begin – the soil. On the one hand, that soil was singularly rich, as rich as the loess soil of the Chinese and the alluvial soil of the Egyptians. It had not their depth, but it had the exceptional contribution of the ash of some 50 craters which are within 20 miles of Rome.

On the other hand, it was placed in a perilous situation if its farmers were to neglect it. It was a wide band or plain, the Campagna, situated between the sea and the steep Alban and Apennine mountains. Upon these mountains rain at certain seasons fell heavily. When there were trees on the slopes, then the rainfall was broken by leaf, twig and branch into a spray before reaching the soil. Where the trees were cut down freely or where the slopes were too steep for them to grow, the torrents of rain reached the earth to beat upon it and send streams of mud sweeping down to the plain. The short rivers between the mountains and the sea loaded with silt. Sometimes their mouths and thus the direct discharge of the water to the sea were blocked, and swamps replaced well-drained land.

Farming in this country, therefore, depended above all on one great feature of farming – proper drainage. Against heavy rain falling upon precipitous hills, the farmers had to protect the soil if they were to be successful. The men of Italy were great farmers, and they accomplished astonishing things.

La Blanchère, through his excavations, revealed in part what the farmers did. He found an extensive engineering system of water control and drainage, including numerous relics of drains, tunnels and dams. Professor Frank writes:

> 'By diverting the rain waters from the eroding mountain gullies into underground channels, the farmers not only checked a large part of the ordinary erosion of the hillside farms, but also saved the space usually sacrificed to the torrent bed. It would be difficult to find another place where labour had been so lavishly expended to preserve the arable soil from erosion.'

Noting the finely trimmed polygonal masonry of the dams, largely made of blocks weighing half a ton each, the professor adds:

> 'It, is impossible, after surveying such elaborate under-takings, to avoid the conclusion that Latium in the sixth century B.C. was cultivated with an intensity that has seldom been equalled anywhere.'

The men of Italy, later to be known after their capital city as Romans, began their unequalled story with a tremendous, vital force – that of an exceptional and well treasured soil. One can immediately realize the vigorous and profound respect for farmers and farming which characterized the Roman poets, prose writers and statesmen of much later ages, and their looking backward to their ancestors as men of exceptional fibre and character derived from their farming. They looked back to something exceptional in seeking the origin of the firm strength of Rome.

These great farmers, who protected their land from the torrential invasions of the climate, had also to protect it against the invasions of human beings – not just neighbours, but also those who had come over the Alps and Apennines in search of land. The farmers then proved themselves great warriors. Farmer and warrior contended within them, but as successes in war grew, so the warrior factor transcended that of the farmer, and the type of farming changed.

The number of small farmers able to support themselves and their families well on less than five acres of intensive farming decreased. From the point of view of the soil, indeed, the story of Rome and its empire was largely a competition between warriors gaining land by conquest and exploitation, and farmers losing it because of enforced, inferior ways of farming, and through erosion. Amongst the splendour of Rome's achievements, this fundamental part of its story has hardly been recognised. In the end, it was the rebellious soil that broke the strength of the warrior.

It is understandable that, if farmers were liable to be called up for national service as warriors, intensive personal farming suffered. The farms could not be kept in good condition when many of the men who worked on them were away at the wars.

This drain began with the wars the Romans fought in or about Italy, but it only became critical at the time of the terrific struggle of Rome against Carthage, and particularly as the result of the fifteen years of Hannibal's warfare within Italy itself. That led to immense destruction, not only of the farmer-warriors themselves, but also of water channels, drainage, farm buildings, roads, bridges, trees, and other essential elements of intensive farming. When the war was over, the government of the victorious but exhausted Romans was faced with the question of the reconstruction of the land.

Now at the same time that this question became paramount in Roman Italy, it also became paramount in China. The Chinese Empire of that time was situated in the middle part of the Huang Ho (known to us as the Yellow River) basin and the great territories on either side of it.

To protect his empire against the warriors of the Tartars, the Emperor, Chin Chi Huangti, resolved to build a huge fortified wall. To build it, he had to procure vast numbers of labourers, and these he had to take from the land. So he abolished the *Tsing Tien* system and the inalienability of the land which was the essential part of it, turned the peasants from their holdings, and sold the land to all able and willing to buy.

In both the Roman case and that of Chin Chi Huangti, the land was the chief source of wealth. The rich men, therefore, readily bought the land of the dispossessed peasant families. So, after the second Punic War in Italy and the building of the Great Wall in China, the rulers of Italy and China were faced with the same question – a question perhaps the most momentous of all in the story of mankind upon the earth: shall the common form of farming be by owners of small holdings, or shall it be that of large estates owned by a small class of wealthy men?

The Chinese chose the former method. After the death of the Enperor revolt broke out, his son was slain and the Han Dynasty (202 B.C.–A.D. 220) brought with it the long struggle between the imperial ministers – who aimed at the restoration of the *Tsing Tien* system of small family holders – and the new aristocracy of large landowners. The struggle was long and bitter, but in the end, except for some large estates which were considered necessary, the *Tsing Tien* system was restored.

This also restored the wisdom of the East, the basis of which was the direct relation of the great majority of Chinese subjects to the soil.

In Italy the same struggle occurred. It was also prolonged and bitter, but here always, albeit slowly, success turned away from small family ownership.

In the peace that followed the conflict with Hannibal, Roman statesmen strove to turn the clock back to the traditional ways of their forefathers, but Rome's conquests and the great influx of foreign slaves to work the land in the place of the dispossessed peasants, in addition to the injury to the soil wrought by the war, weighed heavily in favour of the wealthy classes.

As in China, the land was the chief source of Roman wealth. There were, at that time, no large manufacturing towns, and little commerce. In the words of Professor Frank,

> 'the ancient world has no record of any state of importance so unconcerned about its commerce as was the Roman Republic'.

On the other hand, working in favour of the small landowners was the firmly rooted belief that those who lived on the land were also the finest warriors, and the chief strength of Rome's military power.

The great Roman writers were fully aware of this. Cato the Censor (234–149 B.C.) staunchly maintained that it was the farmers and tillers of the soil who made the best citizens and bravest soldiers.

Varro (116–27 B.C.) voiced the same conviction that country life in its form of peasant-farming was the chief strength of the State.

Cicero eulogized the farmer-citizens who left the plough to save the State, and used his unequalled art to protect the working farmers whose extinction was threatened by the growth of wealthy proprietors.

Virgil used his poetry to exalt the culture of the land and those whose hands produced it.

Horace also described the older type of farming as the best.

Columella, at the time of the Emperors Claudius and Nero (41–68 A.D.), declaimed against the poverty of the land which resulted from handing its cultivation over 'to the unreasoning management of ignorant and unskilful slaves'.

Pliny the Elder, who wrote about the same time as Columella, championed those who worked their own land against the owners of the *latifundia,* or great estates, who abandoned the work to slaves and only kept their country houses so they could hold house parties for their friends. How was it, he asked, that there was so great a fertility of the soil in the past that seven *jugera* (a little over four acres) were held to be sufficient for a farmer and his family? His answer was that in those days, the lands were tilled by the hands of generals and soldiers. Pliny the Elder asked:

> '...whether it was that they tended the seed with the same care that they had displayed in the conduct of wars, and manifested the same diligent attention to their fields that they had done in the arrangement of their camp, or whether it is that under the hand of honest men, everything prospers the better by being attended to with scrupulous exactness?'

This conception lasted up to the time of Vegetius, in the fourth century, who bitterly regretted the abandonment of the ancestral ways, when he saw the poor quality of the military recruits.

The great Roman writers of the latter part of the Republic and the early part of the Empire, then, had a passion and a hope for the reconstruction of the family ownership of the land not only because the farmers were the healthiest, most honest, and most diligent members of the State, but because in times of danger they made the best soldiers.

The military leaders of the late republic were equally convinced of the value in character and physique of the farming class. When the supply of farmer-warriors failed, there seemed to be only one alternative and that was to start with warriors and, as a reward for their services, to give them land to farm.

Marius was the first to make the transition from farmer-warriors to warrior-farmers. He overthrew the tradition that only the propertied classes were worthy to fight for their country. He enlisted the proletariat, especially those who were living on the land, and rewarded their services with a gift of land. Slaves were never enlisted; their grievances were too great, and their numbers too many, for any Roman to dare or even dream of such a dangerous experiment.

When the change to empire brought its long years of peace, it brought with it good government, roads, reliable civil servants, self-governing city states served by an unequalled zeal on the part of public-minded citizens, greater humanity towards slaves, and great prosperity. Of the best part of these first two centuries (from the death of Domitian in 96 A.D. to the ascension of Commodus in 180 A.D.), Gibbon wrote:

> 'If a man were called to fix the period in the history of the world during which the condition of the human race was most happy and prosperous, he could without hesitation name the period between these two dates.'

Even then, the emperors, almost without exception, strove to revive the small family holdings. Augustus and his successors planted colonists on the land; Nerva spent millions in purchasing land for small farmers; generous laws dealt with the food of the agricultural classes; veterans were given free allotments; and Pertinax allowed squatters to occupy uncultivated fields even upon imperial estates, and to possess full ownership if they brought them into cultivation.

Nevertheless, in spite of these desperate endeavours to reconstruct personal farming, the power of money prevailed. The small farming class continuously and literally lost ground, and the wealthy class just as continuously gained it.

In the place of the generous laws of the first two centuries of the empire, there came the restrictive laws of the last two centuries. Agricultural slaves were bound to the land. Heavy impositions and innumerable duties were loaded upon the large class of *curiales*, or members of the senates of the city-states and large villages.

This class of *curiales* included the landowners. As the demands of revenue became more exacting, membership of the *curiae* was made hereditary. The *curiales*, harassed by innumerable officials, duties that could not be fulfilled, poverty which withheld money from the land and forced them more and more to exploit their deteriorating soils, sought by every means to escape from their ruinous property and its duties.

Abbott and Johnson in *Municipal Administration in the Roman Empire* wrote:

> 'Many of them abandoned their property and fled. Others sought to enter some vocation which would give them exemption from municipal charges. The emperors strove to check this movement by binding the *curiales* to their place of origin, and by forbidding them to enter any of the privileged professions.'

These measures failing, laws were then passed under which the property of *all* the *curiae* was made liable for the accumulated dues, the burden then falling on the less fortunate owners. Failure to pay led to the confiscation of property and its transference to the imperial estates which rapidly increased in all parts of the empire, and the tenants of which were exempted from municipal liabilities.

Some also fell to the owners of the great *latifundia*, who were strong enough to either resist the demands of the tax-gatherer or hand on the burden of taxation to their tenants, who had originally sought their patronage as the only way of escape. The *coloni* or voluntary tenants were also bound to the soil, and in the fourth century were reduced almost to the level of agricultural slaves.

> 'The only class in the municipalities not affected by imperial legislation was the proletariat. The practice of Rome in maintaining this parasitic element by private charity was unfortunately widely copied, and imposed a serious charge on the civic budget. Not only that but the glamour of ancient urban life attracted labour from the farms and other industries where a bare living was gained by arduous toil. In the city one could be fed at the expense of the State, and when the *capitatio plebeia* (a tax imposed by Diocletian on the working power of a man in good health) was removed from the residents of the towns, we cannot wonder that the urban movement went on apace.' (Abbott and Johnson).

This downward spiral was both accompanied and caused by the continuous depletion of soil fertility. To this Italy, the imperial mother-country, was the most exposed, and upon her soil its effect was the strongest.

In the early days of Rome seven *jugera* (4¼ acres) were found sufficient for a family, and this was the original assignment given to the *coloni* as tenants of the state. Gracchus, however, found it necessary to increase the assignments to 30 *jugera*.

The fall in fertility due to the war against Hannibal forced upon much Italian land the necessity of large ranches devoted to the raising and feeding of domestic animals or to orchards. This necessity justified the economic brutality of the 'enclosures' of that time, under which land that had previously grown good crops of grain was taken from small farmers by the wealthy classes and turned into ranches.

This, in its turn, confirmed the dependence of the masses on imported corn.

Caesar, as evidence of the soil's further depletion, raised the assignments to 60 *jugera*, and Columella, writing about 60 A.D., asserted that a fourfold return of grain was unknown on Italian farms.

Finally, in the third and fourth centuries, the debasement of the soil completed itself. Much of Italy, once the parent of the sturdy strength of the Latin fathers, became a pestilential swamp. Provinces which had once been the native land of formidable legions were almost bereft of humanity. Flourishing towns dwindled to villages and disappeared. The proletariat of Rome ceased to exist.

The capitalists of Rome did not await the complete degradation of Italy. They transferred their capital at the call of Constantine the Great (288–337 A.D.) to a new capital city on the shores of the Bosphorus, a city situated midway between the rich wheat lands that ringed the Black Sea and the inexhaustible fertility brought annually in the Nile flood. Abandoned Italy fell to Odoacer in 476 A.D.

*

This story will be fraught with meaning to those conversant with – or by a perusal of these pages, about to become conversant with – the past story of agriculture in England and the present state of agriculture throughout the British Empire and other countries of the West.

Amongst other things, they will see the perilous significance of the attempt by the Nazis to conquer the world and bind subject peoples to slavery upon the land. This subjection of the land, against which so many of the great Romans vainly strove, represents the steady and irresistible march towards collapse of a civilization which values the soil as merely a money-making commodity, and not as the very source and creator of the life and health of man.

3

The Roman Foods

IN THE PREVIOUS CHAPTER, we did not address in depth the point that it was the intensive family-based agriculture in a favourable soil and climate which gave the early Romans their physiological vigour and virile character.

We have not proved the proposition, because it is not capable of proof on its own. It can only be presented as a reasonable supposition; that the quality of the food and the health of the individual that eats that food are related. All that we have been able to do, therefore, has been to offer certain facts bearing upon early Italy which will offer some explanation of the exceptional character of its inhabitants.

Let us now review the foods themselves from which this physiology derived. At the outset, we must accept that in history it is difficult to establish the quality and character of the food of a people, and in this the early Romans are no exception.

Dr K. Hintze, however, has in his invaluable *Geographie und Geschichte der Ernährung* collected such knowledge as his persistent scholarship could reveal. What Hintze is able to tell us about the foods of the early Romans is not copious, but nevertheless it is fully in accordance with the diets of some of the most virile people in the world today.

In the previous chapter, we saw what great care was given to the cultivation of food. That is of primary importance. One may presume that with such skilled and laborious cultivation, the soil, itself endowed with excellent natural gifts, provided healthy and well-growing vegetable and animal food. There is no contemporary information, says Hintze, about the foods of early Italy; there are only the traditions, supported by modern research, of what it had been.

Of grains, there was barley, wheat (emmer) and millet. There were no mills, but the grains were crushed in a mortar and the husk removed. The grain was then made into a porridge and eaten with salt. The grains were often lightly roasted so as to make the removal of the husk easier. Later came the hand mill, in which the grain was crushed between two millstones.

The student of nutrition and dietetics will at once note that *only the husk* was removed. The porridge was thus wholemeal porridge and, if flat cakes of bread were made, they too were wholemeal. This traditional porridge, Hintze surmises, was the staple food of all early Romans, who ate alike, as there was little or no food distinction between the classes at that time.

Then came vegetables and fruit. There were cattle, but flesh was seldom eaten, except on the days of religious festivals. The animals were kept for work upon the farm, for the provision of manure, and for milk and cheese, which were an important part of the diet.

The grape was cultivated in Italy in pre-Roman times, but in early Roman agriculture it seems to have played no part. Its culture, however, did reach Italy at some date, and the inhabitants then drank wine. Whether they drank wine made from other juices, as was the habit of the later Romans, is unknown. Barley beer, the drink of northern peoples, never found favour.

The food of the early Italian farmers was, therefore, lacto-vegetarian, a diet which has won high praise from Sir Robert McCarrison and other distinguished modern nutritionists as being the food of many of the healthiest and strongest peoples of the present day. If a healthy soil can be assumed, then the early Romans had in their food all the necessary elements of physiological excellence. (The lacto-vegetarian diet is not the only healthy whole diet. There are other such diets – that of the Eskimos for example, in which the eating of the whole carcass of animals plays almost as prominent a part as it does in the diet of the beasts of prey.)

The lacto-vegetarian diet of wholemeal grains, fruits, vegetables, and milk and its products, as McCarrison has shown, is the basis of the excellent health and physique of the Hunza, the Pathans and the Sikhs of North-western India and, with a more precarious supply of grain and vegetables, of the Arabs and Baggaras.

What proportion of milk and its products was added to the vegetables and fruit eaten by the early Romans is not known. Their value was, one would think, firmly established among the population, some of whose ancestors came from central and eastern central Europe. It was certainly a tradition handed down to and maintained from the early days of the republic.

The *latifundia,* or large estates, of the later republic largely specialized in milk and milk products, as well as wines and olives, and left the growing of corn largely to the provinces. They raised cows, sheep, goats, horses and asses, and the milk and cheese of the milks of all these animals were consumed.

Cossinius discusses the qualities and differences of these products just as connoisseurs discuss wine. Nothing perhaps shows more vividly the immense gap that exists between the sophisticated urban diets of today and that of early and middle republican Rome than this serious devotion to milk and its byproducts.

It is in this lacto-vegetarian character that the early Roman diet is similar, as has been said, to that of many of the healthiest people of the present day. It is in their intensive cultivation of the land through family-based farming that they resembled the Chinese, Koreans and pre-modern Japanese. It is in their traditional reverence for the nutritional qualities of milk and its products, however, that they differ from these far-eastern peoples, whose land supports so numerous a population that there is not sufficient to support a large number of domesticated animals as well.

It is in the combination of the two – intensive cultivation and the culture of dairy products – that the Roman diet most resembled that of the Hunza people of the western Himalayas, who are probably unsurpassed in physique and health by any other people of the present time. Moreover, certainly in mid and later republican times, and therefore possibly in the early Roman period, a great quantity of different fruits were cultivated in Italy, so that Hintze declares that

> '...at Varro's time, all Italy resembled a fruit garden.'

In this generous provision of fruit, the early Roman diet resembled that of the present-day Hunza, who eat great quantities of fresh and sun-dried fruits. It also included great quantities of dates, which those other people of superb physique, the Arabs of Arabia, eat.

As regards early Roman agriculture, the intensity of which has already been described, Frank praises its practical efficiency. Professor Whitney, in *Soil and Civilization* (1926), describes the Roman knowledge of such principles and practices as:

- recognition of the different types of soil and the crops suitable for them;
- recognition of the need for local knowledge of the soil;

- preservation of the soil by successive generations of families, farming in areas where they themselves were born and bred;
- use of legumes, similar to that seen in the agricultural history of the Chinese:
- avoidance of any waste upon the farm, all animal and vegetable refuse being returned to the soil as manure;

and other technical features of agricultural practice which a competent student of practical agriculture such as Whitney is qualified to write about. I wholeheartedly refer the interested reader to his work.

There is therefore, I think, quite sufficient evidence to presume that the Romans and their neighbours belonged to those people who by long adaptation to a repetitive, well-cultivated, sound diet, seem to have acquired an absolute harmony with their food, and were themselves a people of exceptionally good physique and health.

The foundations of their domination of the western world included their diet and their agriculture; and paradoxically, the changes in both that played such a large part in the downfall of Rome came with the spread of that very dominion.

*

The change among rural Italians was much slower in its progress than it was among the rapidly increasing urban populations. The rural people were of course affected by the changes described in the previous chapter, but their foods were still locally produced; milk and its products, grains, vegetables, fruit, oil, wine and occasional meat.

It was upon Rome and other major urban centres that the chief effect was felt.

The bread or porridge of the lower classes was now

prepared, not from local grains, but from grain imported across the seas from Egypt and northern Africa. Emperor Tiberius said to the Senate:

> 'The sustenance of the Roman people is day by day being tossed about at the caprice of wave and storm.'

But that is almost all that can be said with accuracy about the urban lower classes and their food. Hintze laments that

> '...unfortunately, as concerns the life of the smaller folk, comprising the mass of the population, we can learn practically nothing from the writers of the time.'

It is a very different story regarding the wealthier classes of later republican and early imperial Rome. Their change from the simplicity of their ancestors to a life of luxury and indulgence were frequent themes of the writers of the time. Taste and the temptations of delicate dishes replaced the satisfaction of robust appetites. Dinner (*cena*), beginning about 3 p.m., became a cult. Individually and socially, it occupied by its duration alone (three or more hours) a considerable part of the day.

Hintze gives a list of the various foods which reached the tables of the empire: milk, cheese, honey, wine, wheat, barley, millet, beans, lentils, peas, cabbage and other leafy vegetables, tubers, beets, turnips, radish, salad, onion, cucumber, celery, mushrooms, truffles, dill, mint, garlic, coriander, mustard, pepper, cardamon, olives, grapes, apples, oranges, lemons, dates, pears, plums, cherries, figs, quinces, apricots, peaches, almonds, walnuts, hazelnuts, fruit-wines of apple, pear, pomegranate, mulberry and other juices, mutton, goat, pig, deer, boar, chamois, antelope, hare, spiced meats, smoked meats, hams, goose, chicken, ortolan, bunting, starling, thrush, dove, peacock, flamingo, guineafowl, fish, mussels, crabs, lobsters and oysters. Beef was not much eaten, the bullock being kept for labour, and the cow for milk.

There was, therefore, a complete change from the ancestral lacto-vegetarian diet to one drawn from all parts of the available world by the fame and wealth of Rome.

The new diet had what has been termed the 'virtue' of variety. Whether the incentive of variety or the adaptation of familiarity is better for individuals cannot be answered. As far as I can tell, the question is one of those which has had little attention paid to it.

One can only repeat facts. This varied diet is essentially one of wealthy urban or urbanized classes, and it entails gradations downwards to the masses of the urban population. Immediately below the upper class, which gets the pick of the food, there is a grade which gets the foods that are in excess of those required by the rich, or those slightly too spoilt for the fastidious palates of the wealthy. So the diet passes downwards, contracts, and changes to that of the lower classes, who, in the case of Rome, depended for their staple food on distant countries.

It is most important, however, to realize that the defects due to poor food are *acquired* defects and therefore they are not, in the commonly accepted view of modern science, inheritable or inherited defects. Any poor Roman, who by wisdom or fortune, received a good diet from conception onwards, would show the better physique and health which that diet ensured. As for the rich, their varied diet, judiciously used, clearly gave opportunity for health and fine bodily quality, for the rich mostly had estates and other means of access to good milk, cheese, oil, fruit, vegetables and corn.

The rural population, like the wealthy, had access to fresh food. The growing of wheat in Italy did not come to an end. Professor Frank writes:

> 'In Nero's day, Egypt sent about five million bushels of wheat to Rome annually, while Africa sent about twice as much. That would suffice for the capital alone, and reveals why cereal culture could be neglected in the vicinity of the city. But the rest of Italy had a population of about fifteen million, and they would require more than 150 million bushels a year. We must conclude therefore that wheat was very extensively and successfully raised during the first century.'

The foods of Rome during the period of its greatest power may then be summed up broadly in four categories:

Firstly, there were the home-produced foods of the Italian countrymen on their small farms. These resembled most closely, of the four groups, the traditional foods of their ancestry. To what degree they did so is impossible to determine, for as H. Stuart Jones says in *Companion to Roman History* (1912):

> 'though there is good evidence in the literature and inscriptions of the early Empire that the small holding was far from extinct in 100 A.D. and later, we know so little of its working that we can only describe the *fundus* of the capitalistic landowner as Cato and Varro picture it.'

Secondly, there were the home-produced foods of the slave-worked *latifundia*. Under the late republic the condition of the slaves was wretched in the extreme; under the empire their lot was gradually ameliorated. Their foods were presumably not the equal of the first group. Moreover, the specialization of the estate limited the number of foods compared to that produced on the general farm.

Thirdly, there was the varied diet of the wealthy classes, comprised of fresh foods from their own or neighbouring farms and estates, fish from the seas and rivers, and luxury foods imported from abroad.

Lastly, there was the food of the lower urban classes. Of this F. Marshall, in Sir John Sandys' *A Companion to Latin Studies* (1921) writes:

> 'a kind of porridge of wheat, like that eaten in early republican times even in imperial times continued to be eaten by the lower classes ... with green vegetables, seldom meat.'

Grain was apparently still consumed wholemeal rather than refined. As to its quality, there is no means of comparing it with the wheat or emmer and other grains of early Italy, but its wholemeal character was certainly preserved. This is about the only fact of importance one can gather from what is known of the food of the urban lower classes. Nothing is known about their access to dairy foods.

*

Summing up, one may assert that compared to the foods produced by the farmers of early Italy, that of the first group approached – but owing to the increasing difficulties of the farmers, cannot have reached – that of the early period.

The food of the second group, of the agricultural slaves, was certainly inferior.

The food of the third group, the wealthy, is less comparable. It is not possible to state, but it is possible to imagine that it produced a greater variety of human qualities. That it also brought with it the deterioration of over-luxurious and intemperate eating is certain. Nevertheless, the daily life of its eaters, their gymnastics, games, and bathing suggests the persistence of a good level of bodily health and physique.

The food of the fourth group, the poorer urban class, was certainly inferior.

*

With the failing of Italian agriculture, there came a degeneration of foods and their quality, and subsequently a corresponding degeneration of those who consumed those foods. When considering the causes of the decline and fall of the Roman Empire, the degeneration of its foods must cerainly be an important, if not the primary, factor.

This suggests that no empire can endure if the agriculture of the motherland deteriorates. The process is naturally a slow one and as such was not realised by the Romans as a whole, but it was certainly recognised by many of its prominent thinkers, writers and politicians.

4

The Roman Family

THE GROUP BY WHICH the farming of early Italy was carried out was the family. A slave of that time was treated as one of the family, and took his or her part in general work and domestic life for the most part without degradation. The family and the cultivated soil were indissolubly connected; the family was pledged and wedded to the land. The very form of marriage, that of monogamy, was dictated by the soil. The farm provided the family group with food, clothing, shelter, fuel, and an overflow of produce for exchange for goods produced by others. It gave security to the children and old people, and that security was continuous, so long as the soil was well-husbanded.

The unique knowledge of the family was that of their farm and all that affected it. To the family, its land with its particularities was as alive and unique as their own lives. The creation of children to continue the family was, therefore, an aspect of their relationship with the soil.

The farming family was inevitably religious, being so bound with the life of the earth, in which it was itself an active participant in the act of creation. In death resides the inevitability of the resurrection of that which, united with the soil, again becomes alive.

Every schoolboy, recalling his Roman history, is familiar with the stern figure of the *pater familias* – the head of the Roman family who preserved the form of that family, punishing any member of it who endangered its corporative existence, and who did not in extreme cases hesitate to inflict death upon his own flesh and blood.

Ordinarily, one may presume, as member of the family he was probably not so totally grim, but the fact that he had those traditional powers showed that the family was cultivated with as great an intensity as the land. His summoning of the family at the beginning of the day to worship the household gods is an example of the way in which life upon the farm was an extension of the religious life of the time.

The family was large or extended, which is the form particularly correlated with the intensive hereditary cultivation of small farms. It consisted of the father and mother, their sons and grandsons with their wives and children, and their unmarried daughters. The men worked upon the land and for the State, and the women worked for the family.

Outside the family, women generally had no recognized place. She inherited her portion of the family land, but that was for her security, and not to give her individual scope for agricultural skill or toil. She was the mother and the housewife, and in all matters was subordinate to the father.

It was he who had the absolute legal right to decide whether a child born to him or in his family should be reared or not. It was he who ordained the death of a defective child, or one threatening the family unit by over-population. Mommsen states:

> 'The maxim was not suggested by indifference to the welfare of the family. On the contrary, the conviction that the founding of a house and the begetting of children were both a moral necessity and a public duty had a deep and earnest hold of the Roman mind.'

But the family had above all to be strong, both in its own composition and as a functioning part of the State. It had to be strong because the proper service of the soil demanded physical strength, and a strong State – one that could successfully defend itself against invaders and aggressive neighbours – had to consist of strong family units. The family was, indeed, the very essence of the State.

Romaine Patterson in *The Nemesis of Nations* (1907), wrote:

> 'Of all Roman institutions, marriage was the most sacred. The family altar, transmitted from one generation to another and holding a fire which had been lit by ancestors who had been dead for centuries, was the central and most impressive fact in the life of a Roman burgess.'

The economy attached to this sanctity of the family has been described as a 'natural economy'. After the Punic Wars, there arose a rival; the 'money economy'.

The new rich, in the main, were new men; the *equites*. The older landed aristocracy, as was to be seen later in other nations, were not a match for the new men. It was the *equites* who made and controlled the money economy in all its various forms. They paid rents, taxes, customs, excise and other duties. They controlled the import of food, the slave trade, and the creation and circulation of money.

The most certain path to the new wealth was the profession of banking. Only exceptional cleverness or luck in speculation built up wealth more rapidly than did banking, and this very speculation was supported by the bankers. Almost all who took out credit fell into the bankers' debt trap. Successful politicians depended upon the backing given to them by the bankers.

Capital, labour and competition, unknown in the natural economy, became commonplace under the money economy. The bankers became indispensable, and eventually the State itself became an exposition of their power and influence; everything hung from them as the staples of the State. Property ownership became concentrated – the tribune Philippas, quoted by Cicero, stated that there were only 2,000 property and landowners in the whole Commonwealth.

The effect upon the institutions of the family and marriage was profound; they both began to lose their meaning. As the sacredness of marriage and the family faded, it was with the women of the upper class – the class which practically monopolized the pens of the great Roman writers from which we get our information – that the change of values is most vividly illustrated.

The Roman matron of the past now became nothing more than a figure of tradition. The object of the new class of fashionable women was the *reverse* of that of the displaced mistress of the home and family. Her desire was to avoid by all possible means the appearance of being matronly. To conceal all appearances of advancing years – to look young, attractive and ripe for adventure – that was the goal of the new women. Their culture was that of beauty, and their scarcely concealed ambition was to occupy themselves with love affairs without fruition.

Perhaps as a form of revenge for the secret desolation of their wifehood and motherhood, they wasted public resources with lavish prodigality. Fashion and beauty cost so much that thousands of slaves throughout the empire were necessary to support them.

Their passion for personal freedom divided them from the few children which they had. The younger folk, on their part, freed themselves from the shackles of parental authority. The authority of the father vanished into the past, along with the role of the mother.

Family elders, once honoured as the storehouse of experience, wisdom, and links with the past, were now cast aside and ignored, branded with the stigma of irrelevance in their old age.

I cannot better substantiate the accuracy of this picture of upper class Roman women than to quote Theodor Mommsen's account in his *History of Rome*. He is describing the time when society had first achieved a high degree of luxury, thanks to the wealth that accrued from the exploitation of Rome's widespread provinces, and the great number of slaves which filled the place in the Roman world that machines were later to fill in the industrial era. He wrote:

> 'Morality and family life were treated as antiquated things amongst the ranks of society. To be poor was not merely the saddest disgrace and the worst crime, but the *only* disgrace and the *only* crime.'

He described the effect upon society women:

> 'Liaisons in the first houses had become so frequent that only an exceptional scandal could make them the subject of special talk; judicial interference seemed now almost ridiculous.
>
> An unparalleled scandal, such as Publius Clodius produced in 61 B.C. at the women's festival in the house of the Pontifex Maximus, although a thousand times worse than the occurrences which 50 years before had led to a series of capital sentences, passed almost without investigation and wholly without punishment.
>
> The watering-place season – in April, when business was suspended and the upper classes congregated in Baiae and Puteoli – derived its chief charm from the relations both licit and illicit which, along with music and song

> and elegant breakfasts on board or on shore, enlivened the gondola voyages. There the ladies held absolute sway; but they were by no means content with this domain ... they also acted as politicians, appeared in party conferences and took part with their money and their intrigues in the wild coterie-proceedings of the time. Childlessness became common, especially amongst the upper classes, and it was held to be the duty of a citizen to keep wealth intact, and therefore not to beget too many children.'

Childlessness, indeed, had further advantages. Men and women who had children were excluded from the joys of society and were omitted from invitations to social gatherings. Seneca (5 B.C.–65 A.D.), himself a man of great wealth, and whose attachment to Stoic philosophy led him, with his colleague Burrus, to the wise and humane government of the first five years of Nero's reign, did not think it ill, in a manner that would have outraged the farmer Romans, to console a mother who had lost her only son by pointing out that she would now be free to enjoy the pleasures and prestige of society.

Nothing could better than this convey the gulf that had formed between the position of the women of the 'natural economy' and the dominant women of the 'money economy'.

It is more useful to regard this great change as an example of relativity than to condemn it on the grounds of morality. The conduct of the first women was relative to the pre-eminence of the soil, that of the second was relative to the pre-eminence of money.

The first economy sought to preserve life and the soil, the second was complicit in its destruction. Just how destructive it was we shall discuss in the next chapter.

5

Soil Erosion in Ancient Rome

THE BEST SUMMARY of this aspect of Roman history which I have read is by Professor Simkhovitch, in *Rome's Fall Reconsidered*, published in the *Political Science Quarterly* of Columbia University, (1916). Simkhovitch began with quotations from the Roman writers Pliny, Horace, Varro, Columella and others, who were fully aware of Rome's progressive degradation at the roots.

The process was a slow and intensifying exhaustion of soil fertility. It was not due to lack of knowledge of good farming, for:

> '...nothing could be more startling than the Roman knowledge of rational and intensive agriculture.'

Nor, I think, could it be said to be due to debt, for debt did not begin its devastating career until the fertility of the soil became impoverished. Debt was not necessary as long as the farming families were able to give their time to intensive cultivation.

The spread of the degradation of the soil was centrifugal from Italy itself outwards. Varro noted abandoned fields in Italy, and two centuries later Columella, about 60 A.D., referred to all Italy as a country where the people would have died of starvation, if it were not for their share of Rome's imported corn.

The Roman armies moved outwards from Italy demanding land; victory gave more land to the farmers; excessive demands again brought exhaustion of fertility; again the armies moved outwards. Simkhovitch wrote:

> 'Province after province was turned by Rome into a desert, for Rome's exactions naturally compelled greater exploitation of the conquered soil and its more rapid exhaustion. Province after province was conquered by Rome to feed the growing proletariat with its corn and to enrich the prosperous with its loot. The devastation of war abroad and at home helped the process along. The only exception to this of spoliation and exhaustion was Egypt, because of the overflow of the Nile. For this reason Egypt played a unique role in the empire. It was the emperor's personal possession, and neither senators nor knights could visit it without special permission, for even a small force, as Tacitus stated, might "block up the plentiful corn country and reduce all Italy to submission".'

Latium, Campania, Sardinia, Sicily, Spain and Northern Africa were treated as granaries, were successively reduced to exhaustion. Abandoned land in Latium and Campania turned into swamps, in Northern Africa into desert. The forest-clad hills were denuded.

G. Jacks in *The Rape of the Earth* wrote:

> 'The decline of the Roman Empire is a story of deforestation, soil exhaustion and erosion. From Spain to Palestine there are no forests left on the Mediterranean littoral; the region is pronouncedly arid instead of having the mild humid character of forest-clad land, and most of its former bounteously rich top-soil is lying at the bottom of the sea.'

The same fate at a later date fell upon Asia Minor, the decline of the East repeating that of the Western Empire. Sir William Ramsay, in *The National Geographical Magazine* of November, 1922, wrote one of those articles which almost stagger one with the importance of the treatment of the soil in the story of mankind.

The Province of Asia...

> 'in Roman times was highly populated and therefore highly cultivated...
>
> It is difficult to give by statistics any conception of the great wealth and the numerous population of Asia Minor in the Roman period.
>
> In the single province of 'Asia', to use the Roman name for the western part of the peninsula, which was the richest and most highly educated of the whole country, there were 230 cities which each struck its own special coinage under its own name and had its own magistrates, each proud of its individuality and character as a self-governing unit in the great Empire.'

Sir William carried out a careful exploration of some of the areas of high cultivation, which he regarded as the necessary basis of this wealthy province. He found what is found elsewhere – hills denuded of forest and swept by heavy seasonal rains. He also found the relics of the extensive terraced engineering, by which the nourishing water had once been conserved and distributed:

> 'In older time, the numerous terraces would have detained the water from point to point up the mountain side, preventing it from ever acquiring a sufficient volume to sweep down in a destroying flood.'

Against this fertile land came invaders. First came the least destructive, the Arabs; least destructive because they observed in war the sanctity of trees. The Arabs could under the rules of war destroy the crops and produce of the enemy, but only exceptionally the trees, which conserved the soil.

‘It was left to the Crusaders under the command of German, Norman and Frankish nobles and bishops to inaugurate the era of total destruction of a country by cutting down the trees... These broke the strength of an organized society by reducing a great part of the country from agricultural to nomadic status.

The supply of food diminished accordingly, and with the waning of the food-supply the population necessarily decreased.

A decreasing population was unable to supply the labour necessary to maintain the old standard of water engineering on which prosperity rested. Gradually industries languished and died in the towns, as did as the agriculture in the country.

The Sultans did what they could. Neither the Seljuk Turks nor the Ottoman Turks were motivated by fanaticism. They wished to preserve the old social system so far as it was consistent with their interests as a conquering caste; but they could not maintain the education which was necessary to the old Roman system.

Thus the whole basis of prosperity was wrecked, not by intention, but by steady decay. A number of causes co-operated, and each cause intensified the others. Can the prosperity of this derelict land be restored?’

6

Farmers and Nomads

The Land

Physical maps, showing the different elevations of land, have always had an irresistible attraction for me, and none is more attractive than that of the vast continent of Asia with its European appendage, pushed out like a tongue between the Mediterranean and Northern Seas.

What a huge playground of history this map presents! There has been nothing like it in the other continents of the world, in Africa, Australia and the Americas. They are, excepting Egypt, almost without known history, compared to the Eurasian continent.

The map which I possess has five colours to denote different heights, dark green to show land below sea level, light green from sea level to 500 feet, yellow 500 to 2,000 feet, light brown 2,000 to 5,000 feet, and dark brown over 5,000 feet.

Belt No. 1

Asia begins with the beaches of the Arctic Ocean. Then comes a vast light green band or belt with a few yellow areas within it. It stretches right across Asia and into Europe. In Asia it is the Siberian Plain; in Europe the Great Lowland Plain. Except for an extreme northern band of Arctic vegetation, the *tundra*, this light green belt is forest land with great rivers passing through it to the Arctic Ocean. It is *Belt No. 1*. It has played very little part in Asiatic history.

Belt No. 2

The land between this green belt and the mountains to the south of it is yellow-tinted. It has less rainfall than the green land north of it, and is subject to seasons of aridity. It is grassland, the land of the steppes. This is *Belt No. 2*. It has played a great part in Asiatic history. Belonging to *Belt No. 2* as steppe land, there is a patch of light green near the Caspian Sea. It is a part of the Kirghiz steppes and it passes directly to the steppes of south-eastern European Russia; when north of the Caspian, it is actually tinted dark green or below sea level. This is the Caspian Tract, through which so many hordes of Asians have passed into Europe in prehistoric and historic times.

Belt No. 3

The third belt begins with light brown almost from the northeastern tip of Asia. It then shows a dark brown series of mountain ranges. From east to west these are the long thin line of the Yablonoi Mountains, the much greater mass of the Sayan and Altai Mountains, and the lofty Tianshan, which ends at the seventieth longitude in the Pamir, or roof of the world.

Belt No. 4

is a light brown belt between *Belts 3* and *5*. It includes Mongolia, the Gobi Desert and Turkestan. It comes to an end at the Pamir. Mongolia is steppe country and its inhabitants have played a large part in history, not only of Asia but also of eastern Europe. The name Mongol, or Tartar, however, has become attached to other peoples of the steppes as well as to the people of Mongolia.

Belt No. 5

The filth belt constitutes the largest mass of elevated land in the world. In the east it rises almost abruptly above the light green of the lowland of China, and then forms the most extensive elevation, that of Tibet, 11,000 feet and over, which is inhabited by man. Tibet's southern border is formed by the highest mountains of the world, the Himalayas. The Himalaya pass on westwards, forming the northern barrier of India and join the lofty Tianshan of the third belt in the Pamir.

From the Pamir the conjoint Tianshan and Himalaya continue westwards as the Hindu Kush Range; thence reaching across northern Afghanistan and Persia to arrive at Ararat in the east of Asia Minor. Ranges of lesser height pass from the Hindu Kush southwards to form the eastern border of Afghanistan and then pass west and north-west to the east of the Persian Gulf as the mountains of west Persia, and so reach Ararat. They, and the northern ranges, enclose a smaller and much lower plateau than that of Tibet, the Iranian Plateau. Finally from Ararat, mountains continue westwards in Asia Minor, and appear in Europe as the Balkans, the Alps and the Pyrenees.

Belt No. 6

is the land of the farmers. For our purposes, it is the light green land about the great rivers; the Huang Ho or Yellow River and Yangtse Kiang of China, the Brahmaputra, Ganges, and Indus of India, and the Euphrates and Tigris of Iraq.

Such, in brief, is the physical map of Asia. Its fascination lies in the fact that one can read from it some of the vast history that has been played out upon the huge stage of the continent of Asia.

The Nomads

The nomads are the inhabitants of *Belt 2*, the steppe country. They are defined in *Annandale's Concise English Dictionary* as

> 'those people whose chief occupation consists of feeding their flocks, and who shift their residence according to the state of the pasture.'

The nomads, according to this definition, present a picture of wandering shepherds and peaceful pastoralists passing from pasture to pasture with their herds. They would erect their tents of ox hide at new pastures, and enjoy the comfort of a home and resting place until their experience told them that the pasture was insufficient for their cattle, and it was time to move on.

Probably in the earliest historical times, the nomads had horses. The horse is an Asiatic animal and the only wild horse now known is found in Western Mongolia, as a natural inhabitant of its dry, open steppes. Certainly the nomads had horses before 2000 B.C., for horses appeared in Babylonia at that time and two centuries or so later the Hyksos, who conquered Egypt, introduced horses into that country. So we can add the horse to the company of nomads.

The horse was to them a noble animal and was ridden only. It was not used as a beast of burden as it is today; it was the oxen who drew the heavy wagons when they trekked. The horse was loved for its speed. It was the swiftest animal of the steppes and it was this which made it loved by the nomads.

This picture of the nomads is a pleasant one and their life was indeed peaceful and pleasant, but only as long as the pasture was good. When the rain was scanty and the pasture poor, they were in trouble.

Then they had to move frequently and, sometimes,

faced by the loss of their cattle by starvation and themselves feeling the pinch of hunger, they would move quickly, and their warriors, mounted on their loved steeds and armed with bows and arrows, would fling themselves upon peaceful people, either pastoralists like themselves or farmers, slaying many and taking possession of their land. With their incredible swiftness on the march and an unprecedented speed of encircling attack, with the deadly accuracy of their arrows shot from the saddle, with their horrific cries terrorizing their victims, they must have seemed like a horde of winged tormenters whose sting was death. Avoiding capture or destruction were impossible.

The cause of this disturbing loss of food was at one time believed to be an increasing dryness of the climate in historical times. This hypothesis was propounded by Prince Kropotkin in an article in the *Royal Geographical Journal* (1904) in which he stated that it was quite certain that *Belt No. 4* was more populated than it is now; it was quite certain, for example, that within historical times, Eastern Turkestan and the adjacent part of Mongolia

> '...were not deserts as they are now. They had a numerous population, advanced in civilization, and which stood in lively intercourse with different parts of Asia'.

Many of them were successful farmers, dependent on irrigation from rivers flowing from their enclosing mountains. This, Sir Aurel Stein, in his monumental work *The Desert Cities of Cathay* (1912), has convincingly proved beyond further discussion. Kropotkin continued:

> 'All this is now gone, and it must have been the rapid desiccation of this region which compelled its inhabitants to rush down to the Jungarian Gate (a name for Western Mongolia), to the lowlands of Balkash and the Obi.'

Huntingdon Ellsworth skilfully developed this hypothesis in *The Pulse of Asia*. The hypothesis gave rise to very widespread investigation, with the result that, though fluctuations of climate undoubtedly occurred, as shown, for example, by the rise and fall of the level of the Caspian Sea, nevertheless a continuous decline in humidity in historical times could not be accepted. Drives through the Jungarian Gate were, however, accepted.

Another reason had to be found. It was found in the treatment of the soil by the nomads.

The first statement of this which I have been able to find is by Monsieur Rorit in the *Royal Geographical Journal* (1870), where he wrote:

> 'The nakedness of Arabia and the vast tracts of Asia in the north and west, the sterility which extends over Persia, cannot be traced to any other cause than the pastoral habits of the inhabitants. The people inhabiting them are locusts; they destroy all woodland and vegetation, modifying even the climate – whence the necessity of migrations. Had the invasions of the barbarians any other cause? A study of the question in this sense would perhaps give us the key to the great migrations of mankind.'

Rorit's reasoning is pungently expressed, but it is correct. The same process is going on in many parts of the world today, before our very eyes.

In the countries in which nomads fed their flocks and herds and grew temporary crops of grain, there was, as is usual in nature, a delicate balance between animal and vegetable life. Animals feed upon the land and manure it, but they do not ravage and destroy it.

When human pastoralists entered these areas, they brought with them an altogether new danger, namely their status as a being so advantaged by their upright position, their hands and their large brains, that they had the capacity to override the natural law of balance.

They could breed more animals than the land could properly support; they could break up the natural life cycle of a district by using all that the soil produced, and then, when exhaustion of the soil came, move on to another district.

With weapons forged from the iron of the Altai Mountains, these nomads could cut down trees and shrubs and, with their ability to create fire, they could burn as well. The ash of burnt trees and shrubs fertilised the land and enabled the nomads to grow good temporary crops for a number of seasons. They, in short, had the power to exceed the limitations set by nature.

Nature followed the rule of return, while the nomads, unlike true farmers, failed to follow it. By cutting down trees and shrubs for fuel and for ash, they eventually made the soil drier.

When rain fell onto the steppes in their normal state, it was broken into a fine spray by trees, shrubs and thick grass, and was thus evenly and widely spread over the topsoil. The topsoil, sheltered from sun and rain, stored the water. By slow evaporation from the vegetation, the water was returned to the air.

But where excess of cattle fed upon the land and where trees and shrubs had been burnt, the soil was exposed, dried and powdered, and then blown away by the winds or washed away by the rain.

So a desert was formed, forcing the nomads to move on. Nature then returned and in many cases restored the ravage. But if the destruction of fertility had been too great, or if the half-recovered soil was again used for crops and grazing, permanent deterioration was the result.

The nomads, then, lived a life of imbalance by not following the rule of return, which is *the only stable rule of living.*

They were forced instead to live a life of *chance*. They depended on the seasons and, as the seasons varied, the nomads were essentially *speculative*. In this character, indeed, they were similar to other kinds of speculators, even those of the present time.

Speculators disregard the rule of return. They strive to gain without giving; they disregard future generations; they are indifferent to the sufferings of others, and care only that they themselves escape suffering. Eventually, though, there is no escape from the effects of their actions, because ultimately their values are destructive, and not conservative.

As long as the nomads failed to use settled agriculture and limit their cattle breeding, life was sometimes generous to them, sometimes even-handed, and sometimes, as in seasons of drought, harsh.

During hard times, the nomads organized wide-sweeping hunts of wild animals for their food. If further pressed, they were forced to move on, and this sometimes entailed making raids into the lands of their neighbours, who, in their turn might themselves take to raiding.

Thus, with increasing numbers, they made themselves masters of the lands of settled farmers, and the food and wealth which they had not the wit to get by their own skill and toil became theirs as well.

Hence they praised and practised war, not as a means of defence in the way in which a sturdy peasantry has so often successfully defended itself and its soil, but as a road to mastery and wealth. To them, life was not only a struggle for existence, but a desire to wield power over their enemies, an assertion of the right of the better-armed over the weaker.

They terrorized when they attacked, and, when they conquered, they were successful owing to the speed of their attack, the terror they aroused, and the human slaughter they inflicted. All these characteristics of theirs ultimately, therefore, arose from their attitude and relation to the soil.

The soil was something to be exploited and even plundered for their gain. This attitude was in sharp contrast to the tenet of the Babylonians – that the soil belonged to their god – or to the sanctity with which the soil was endowed by the followers of Zoroaster. These faiths of the holiness or wholeness of the soil were, as we shall see, faiths of the farmers; the very word *cultivate* is derived from the Latin verb *colere*, which has a two-fold meaning – *tilling* and *worship.*

Yet the nomads were not always wild horsemen, as they were when they went to war against the farmers. They also had within them the kinder apects of humanity. Professor Keane, in *Man, Past and Present*, said of the Tartars, or Mongols:

> 'They are brave, warlike, even fierce, and capable of great atrocities, though not normally cruel.'

The invention of the gun has now robbed them of their power and, in consequence, they have

> '...almost everywhere undergone a marked change from a rude and ferocious to a milder and more humane disposition.'

The nomads have been the great human desert-makers, and the deserts of the Gobi, the Lop Nor, the Taklamakan, the Registan, the Great Salt Desert, the Syrian Desert, and even the Arabian Desert and the Sahara of Africa are all results of their treatment of the soil.

Nor is this desert-making by men at an end. It is going on at the present, as future chapters will show, in North and South America, in Russia, in Asia, in North and South Africa, in Australia, and even in the islands of New

Zealand and the West Indies, with a speed that outstrips that of the Asian nomads, so much so that it may even be said that man, in this proud scientific era, has paid for his all-too-swift advances with the loss of his most valuable asset – the fertility of the soil.

He has become the great transferrer of the wealth of the fields, and, by his destruction of the soil, humanity has condemned itself to God knows what impending calamities, dwarfing those brought about by the Asian nomads – unless we call a halt. It is this fact which gives this dissertation on the nomadic character its significance and urgency.

The Farmers

Belt No. 6 of Asia is the belt of the farmers. From the mountains of *Belt No. 5* great rivers run southwards into the Persian Gulf, the Indian Ocean and the Pacific Ocean. Along these rivers the farmers built up their civilizations.

The first civilization we shall consider will be the one that is believed to be the oldest of them, with the possible exception of China. It is that of Iraq. This civilization was, *par excellence*, the civilization of irrigated farming.

It was centered around three rivers – the Euphrates, the Tigris and the Karun. All three rivers, during this epoch, discharged their waters separately into the head of the Persian Gulf. Today the Karun joins the Tigris, and the Euphrates and Tigris have one joint mouth 140 miles to the south of where the three rivers once met the sea.

The first river of the three to support a civilization was the shortest and the most eastern, the Karun, with its important tributary, the Ab-I-Diz. These two rivers ran through flat alluvial country before they reached the sea. Their courses in the flatland were brief compared to those of the Euphrates and the Tigris, their major lengths being through the mountains. Elam, as this country was named, therefore resembled Italy in having a plain near the sea, and a great capital, Susa, situated on the plain within 30 miles of the hills.

The civilization of the farmers and hillsmen of Elam preceded that of Italy by some 3,000 years. Elam showed much of the tenacity of Rome, for it largely kept its independence and played a considerable part in the riverine civilization of Iraq for some 2,000 years.

The riverine civilization was further developed by Sumer, Akkad, and Babylon, all three city states being watered by the Euphrates. The Tigris was swifter and more steeply banked, and therefore less used.

The Akkadians and Babylonians were Semitic. The Sumerians were of doubtful origin. They were believed to have preceded the Semites, and to have been the inventors, about 3500 B.C., of the cuneiform writing later adopted by the Semites and found upon baked clay tablets, the excavation and deciphering of which have enabled scholars to extract from the sites of the city states the history of this artistic, flourishing, powerful and very ancient civilization based on irrigated farming.

The city states consisted of the cities themselves and the cattle pastures, together surrounded by walls, and also of the farmed land outside the walls. The life of the land depended solely on irrigation and it was the ambition of good rulers of the city states to cut out new canals and clean out the old ones. The early history of the tablets records such work, the building of temples and the wars carried out by the cities against each other. The purpose of the wars was to establish suzerainty, but not in any way to injure the farming of the soil, upon which all depended for their existence.

Eventually Babylon became paramount. Babylon's

first dynasty began around 2400 B.C., and it was finally conquered by the Persians under Cyrus in 538 B.C.

Lastly in the sequence of these riverine civilizations came the Assyrians, also a Semitic people, appearing in the thirteenth century. They inhabited the land of the middle reaches of the Tigris.

From the level of Hit on the Euphrates, a little to the north of the modern Bagdad on the Tigris, the land for 550 miles to the Persian Gulf is purely alluvial, with all the advantages of alluvial soil such as lower Egypt enjoys from the Nile, Bengal from the Ganges and the Brahmaputra, and the Chinese in the lower reaches of the Huang Ho and Yangtse Kiang. Above Hit there is a reef of hard rock, from which to the north the land continues to be rocky.

For this reason the Assyrians, with their capital at Kalaat Shirgat on the Tigris about 200 miles to the north of Babylon, were not so favoured as the southern alluvial peoples, and therefore exhibited what Sir Percy Sykes, in his *History of Persia*, calls a 'predatory character'. Their initial strength, says Sykes, lay in the formidable fighting quality of a free agricultural class.

When this class became exhausted, the Assyrian rulers moved their capital to Nineveh on the opposite bank to modern Mosul, near where the Tigris enters Iraq from the mountains of the south-eastern corner of Asia Minor. This gave them the control and use of the hillsmen as mercenaries.

The Assyrians, who were northerners, became masters of the southerners of Babylon in 745 B.C., and remained so until 606 B.C. – a brief period of about a century and a half, a pattern which is common in the case of inferior conquerors. In 606 B.C. the Medes, with the assistance of the rebelling Babylonians, sacked Nineveh. So great had been the cruelty and barbarity of the northerners compared to the southerners of Babylon, that Sykes declares:

> 'Assyria shone only as a great predatory power, and when she fell, passed away into utter and well-merited oblivion'.

Assyria's predatory character introduces us to the Aryans, for the Medes were Aryans, living in the valleys of the Zagros Mountains, and the adjacent Iranian plateau in the north-west of Persia.

The Aryans entered the north of Persia about 2400 B.C., and the Medes about 2000 B.C. They were steppe dwellers, as their language, in its omission to speak of forests and mountains, demonstrates. They came as nomads with flocks and herds, moving their habitation from place to place.

The Medes, at the time of Assyria's ascendancy, were subject to predatory raids by the Assyrians. The results of these raids show the Medes as a more settled people than were their nomadic ancestors. Sykes writes:

> 'From the frequency with which these expeditions raided the Iranian plateau – the plateau that is today so desolate – and from the number of towns they destroyed, we can infer that it was then a distinctly fertile and well-populated country. The inference is confirmed by the number of prisoners and the thousands of horses, cattle and sheep that were captured. Thus in one raid in 744 B.C., the success of the campaign may be estimated from the fact that 60,500 prisoners and enormous herds of oxen, sheep, mules and dromedaries were led back in triumph to Calah, near Nineveh.'

These afflictions brought about a desire for vengeance among the Medes. They were sturdy hillsmen and unexcelled horsemen. Under Cyaxares, their great leader, they circled round the Assyrian army just beyond the range of their weapons, and poured a ceaseless shower of arrows into their midst. With the help of the Babylonians, they destroyed the Assyrian Empire.

The Persians entered eastern Persia from the steppes to the north of Khorasan in what is now Russian Turkestan and, traversing the south-eastern Persian province of Kerman, reached Fars, with the Persian Gulf as its western limit and Elam and the Medes to the north. At this time a notable event happened, which illustrates the soil-based character of the Medes and Persians.

They both adopted the religion of Zoroaster, who was born about 660 B.C. (or perhaps a few generations earlier), and therefore some half a century or more before the destruction of the Assyrians.

Zoroaster raised the use of the soil to the first place in the three chief tenets of his religion. His first tenet was that *agriculture and cattle-breeding are noble callings*:

> 'He who sows the ground with care and diligence acquires a greater stock of religious merit than he would gain by the repetition of ten thousand prayers.'

A further illustration of the character of these farming and pastoral peoples in the highlands of Western Persia is shown by the remarkable fact that Cyaxares and his Medes did not take possession of the wonderful riverine civilizations of Iraq after the sack of Nineveh. They were content to hand it over to their allies, the Babylonians, who then erected the brief but brilliant Tenth Dynasty. Cyaxares, however, did not cease from his conquests, but confined them to the uplands of Persia, Armenia, the upper reaches of the Tigris and western Cappadocia.

One of the greatest of Aryan leaders, the Persian Cyrus, defeated the son of Cyaxares by taking Ecbatana, the modern Hamadan, and the capital of the Medes, in 550 B.C.

Cyrus became the first king of all Persia, and proceeded

to make himself master of the most extensive empire the world had then seen. From 500 B.C. to 600 A.D., Persia controlled an area more than half the size of Europe. The Medes were not made the subjects of Cyrus, but were his brethren in religion and status. He overthrew Croesus in 546 B.C. and became master of the Greek colonies in Asia Minor. He took Babylon in 538 B.C.

So the early Persian conquests, it seems, were not based on the strength in food of Iraq. By becoming the master of Iraq, Cyrus brought the independence of its riverine civilizations to an end.

Now, this long story of the riverine civilizations, enduring as it did for 30 centuries and surpassed only by the 40 century history of China, illustrates the *extraordinary stability of a civilization founded upon the soil as its first principle.*

The city states of Babylon regarded the land as sacred. Each state had its god who was the owner of all the city land – its *belu*, or 'Lord'. The priests acted as his agents. Sykes defines it as a feudal, ecclesiastical system, but the fact remains that *the soil was regarded as sacred.* This sanctity was revived by the followers of Mohammed when they became masters of Iraq.

A second notable fact of the 30 centuries of the civilization founded on farming was *its freedom from destruction by the nomads.*

Only once did nomads threaten its independence; the invasion of Iraq by the Semitic Aramaean hordes from Arabia. They took the whole of Assyria, and brought the Eighth Dynasty of Babylonia to its end. They were eventually subdued by the Assyrians. With this exception, as well as a brief raid by Scythians sent against the Medes by the failing Assyrians, the invasions and conquests of farmers by nomads, which have played so large a part in history generally, did not greatly affect the strength of the organized societies of the farmers.

It is to the vast effects on history that these invasions have had that we now turn.

Nomadic migrations and farmers

The first great migrations of the nomads occurred between 2500 and 2000 B.C. During that time the Aryans, as we have seen, reached the Iranian Plateau. It was the time of the first and second dynasties of Babylon, and apparently had no effect upon the highly organized riverine civilizations.

The second migratory period was about 1500 B.C. It was the time of the overthrow of the early Minoan civilization of Crete by the Dorians, the conquest of Egypt by the Hyksos and the disturbances of the first dynasty of China, the Shang, (1750–1122 B.C.), by the Mongolians.

The third migratory period occurred about 1200 B.C. It was the time of the invasion of Greece by the Dorians and their destruction of the Minoan civilization, and the end of the Shang Dynasty of China at the hands of the Mongolians.

Neither of these two periods of nomadic migration affected the riverine civilization of Babylonia. The Kassites, who formed the third dynasty of Babylonia (1700–1170 B.C.) and came from the Zagros Mountains, though originally perhaps nomadic Aryans, were not at this time nomadic, but a settled people like those of Elam, their southern neighbours amidst the hills.

The fourth migratory period witnessed the virtual fall of the Chow Dynasty in China in 659 B.C. The same movement brought the Sesunaga to India in 620 B.C. They established the Magadha Kingdom in the central and eastern Gangetic Plain. Possibly contemporaneous movements in Europe were those of the Celts into the middle valley of the Danube, and from there at a later date into France, Spain and Northern Italy.

The remarkable fact, then, about these four great migratory periods of the nomads is that they had little or no effect upon the first and perhaps greatest Asian farming civilization, though they were so destructive to other lands and peoples.

The Persians, who succeeded the Babylonians, were no less strong. They were Zoroastrians, and we have already seen the importance that Zoroaster attributed to farming.

From the time of Cyrus and for a long period later, Persia offered an almost invincible obstacle to the movements of the nomads of Asia, diverting them to India to the south and northwards to Central Turkestan and Europe. Persia fell to Alexander of Macedon, and after his death the Seleucids reigned. They were replaced by their pupils the Parthians of Khorasan, like the Persians an Iranian people, the words *Arya* and *Iran* having the same derivation.

The Parthians, in their turn, gave way to the Persian Sassanian Dynasty, and the Sassanians in turn to a people who rapidly became great farmers – the Arabs of Islam.

The significance of this barrier of farming civilizations against nomads is very great indeed. It began with Babylon's first dynasty, around 2400 B.C., and it endured until the overthrow of the Arabs by the nomadic Mongols in 1258 A.D. – a total of some 4,000 years.

The next great farming people of Asia were the Chinese, and they can also claim a history of 4,000 years. Professor F. King, who quite recently wrote his famous book on their agriculture, called it quite correctly *Farmers of Forty Centuries*.

The first location of the Chinese was along the Huang Ho or Yellow River, which arises in the highlands of Tibet, as does their second great river, the Yangtse Kiang. They settled upon the lands along the Huang Ho after it makes its right-angled bend from east to south at the fortieth latitude.

The nomads, who so frequently threatened this otherwise peaceful people, were called by them the Tartars, and by us the Mongols, and their country Mongolia.

The Chinese are historically minded. They begin their history with the Emperor Fuhi (2852–2738 B.C.), who is said to have founded the patriarchal family system. During the reign of the Emperor Huang-ti (2704 or 2491 B.C.), the northern Mongols receive their first mention under the name of Hun-yu. The date of the first definite dynasty, the Shang, is given as 1766–1122 B.C. During their time, Chinese history was mainly one of peace, but towards the end of their period, the Mongols, known now as Hiung-wu, appeared, and it is said that it was with their help that the Shang Dynasty was overthrown by its successor, the Chow (1122–659 B.C.).

The Chow and the Shang Dynasties together reigned for a thousand years. In this length of time, they resembled the Babylonians. When the Chow Dynasty came to an end, general unity ended with it. Various states, especially the border states, asserted their independence and fought together for suzerainty. It was in this period of 'Contending States', as Chinese historians call it, that, from 551 to 479 B.C., Confucius lived.

Another very famous ruler changed the complexion of the period of 'Contending States'. He was the Emperor Chin Chi Huang-ti, who ruled from 249–210 B.C. This Emperor united the Chinese, and to prevent the invasions of their troublesome neighbours, the Mongols, he built perhaps the most prodigious Maginot Line mankind has ever witnessed. The immense fortified Great Wall stretched from the sea to the north of Beijing for 1,500 miles. Nor did the Emperor's energy exhaust itself in this huge undertaking.

He drove the Mongols out of Inner Mongolia, on the borders of China, into Outer Mongolia, and the Early Han Dynasty (206 B.C.–23 A.D.) continued the aggression. Then occurred one of those remarkable population movements which the Chinese, in an aggressive imperial mood, originated.

The Mongols retreated westwards, forcing other peoples before them. Some of these peoples, continuing westwards, conquered the Greek kingdom of Bactria-Sogdiana between the Hindu Kush and the Sea of Aral; others turned south and eventually passed through the Bolan Pass and invaded the Punjab.

The early Hans annexed Mongolia and Eastern Turkestan, and Bactria and Sogdiana were compelled to acknowledge their supremacy. Any further conquest was then stopped by the barrier of organized Persia.

The Hans, early and late (23–230 A.D.), bring us to the Mongol or nomadic movements of our own era. A great Mongol movement brought about the downfall of the Western Tsin Dynasty in China in 419 A.D. The Gupta Dynasty in India was overcome by the White Huns in 450 A.D. These White Huns also for many years harassed the Persians, but were eventually destroyed by the power of the Sassanian Dynasty. The date of the movement under Attila the Hun, who reached Rome, is given as 445–453 A.D., and caused the Slavs to push the Teutons into Britain, France, Austria and Lombardy.

The Chinese, under the short Suy Dynasty (590–618 A.D.), and the Tang Dynasty (618–907 A.D.) in its early period, launched an imperial recoil movement and by 640 A.D. had again conquered Eastern Turkestan and extended their influence as far as Persia and the Caspian. The Arabs in Persia checked their threat to Persia about 650 A.D.

A further Mongol movement brought the Tang Dynasty to an end in 907 A.D. and sent the Turki-Mongol Ghazni Dynasty into North-western India. The Magyars entered Europe and divided the Slavs into northern and southern Slavs.

The vast Mongol movements under Genghiz Khan and his successors occupied the thirteenth century. China and Northern India were conquered. Arab power was broken in 1258 A.D. Assaults were made on the Byzantine Empire, and southern Russia was conquered and occupied.

The latter half of the fourteenth century witnessed the peculiarly personal achievements of the greatest of Asiatic conquerors, Tamerlane (1335–1405 A.D.), the Turk. No ruler of the time was able to oppose his supreme genius. He was not destructive and murderous as were Genghis Khan and his successors. He was, wrote Sykes,

> 'profoundly sagacious, generous, experienced and persevering. In *The Institutes* it is said that every soldier surrendering should be treated with honour and regard, a rule which, in striking contrast with the customs prevailing at the period, is remarkable for its humane spirit.'

As a consequence, no great movements of peoples occurred during his reign.

The last two Mongol movements were conquests and changes of dynasty without any general effect. The first was the Moghul conquest of most of India (Akbar, 1556–1605 A.D.); the second that of the Manchu conquest of China in 1644 A.D.

*

Such, in outline, was the historical effect of the nomads of Asia. A full account of these movements, the history caused by them, and the numerous dynasties founded by them in Asia and in Europe, mostly to endure only about a century and a half, can be found in my *Causes of Peace and War* (Heinemann, 1926).

Here let us close the physical atlas, and this long chapter, which it is hoped has demonstrated to readers that the origin of much war and history lies in men's attitude to the soil.

At present it is a subject mostly ignored by the historians, but I hope that soon some prominent modern scholar will deal with the subject more adequately than I have been able to do. Perhaps it will give rise to a greater knowledge of the causes of devastating wars and their prevention.

Perhaps also it will show that family-based farming establishes in each nation a class strong in its desire for peace, and that as in the past, the nomads were the enemies of peace. The nomadic type is still prevalent and powerful, and still sees in war the means of its advantage. The social elimination of these advantages and a true valuation of the soil may yet prove to be powerful factors in the maintenance of peace.

7

Contrasting Pictures

IN ORDER TO GET a clear idea of the modern perception of the soil and its effects, we will begin with the opposite of the unavoidable sketchiness of a transcontinental survey such as that of the last chapter, and we will now concentrate on self-contained examples, on a smaller scale.

Small islands offer themselves at once as the opposite to great continents. By nature they are self-contained. Their inhabitants get food from the sea, a source with which they are unable to interfere, as they can with the soil.

Sea food has the natural quality of wholeness; health, therefore, should be found in such islands. It is never vain to ask health from nature, and small islands still preserve to a large extent this gift – or certainly they did until trade intervened.

The health of the few inhabitants of an isolated island, Tristan da Cunha, was described by a medical visitor in the *British Medical Journal*, March 1938, as being vastly superior to that of the civilized world. Similar health distinguished the inhabitants of once isolated Iceland and the Faroe Islands. The health of the South Sea islanders provides further examples.

Such good health is shared by all forms of life and is preserved as long as the island is a self-contained life cycle and the rule of return is obeyed. Once trade enters and breaks the rule of return, however, deterioration of life sets in.

This has happened in the Falkland Islands in the South

Atlantic. Here there are many islands, but only two are of any size, being some 100 miles long. These islands came into the hands of the British over a century ago. Bleak and almost treeless, they nevertheless possessed vegetation upon which sheep and cattle could feed. These animals were bred for the British market.

The venture was a success, but over time it became clear that the way in which the trade in animals was being conducted was inimical to the life-creating soil of the islands. Sir John Orr, in *Minerals in Pasture,* states:

> '...in the Falkland Islands, sheep have been reared and exported for 40 years without any return to the soil to replace the minerals removed. During the last 20 years it has become increasingly difficult to rear lambs. The other animals are also deteriorating.'

The sequence, then, is a simple one. The two islands had their own life cycle. They became British possessions, and sheep and other domestic animals were imported into the islands. These animals were not predatory; they are grazing animals, which all over the world feed on grass and other herbage, and do the soil no harm.

Presumably, then, they could have been added to and supported in the Falkland life cycle without doing it any harm, if they had done their grazing and at the same time (and finally after death) returned to the soil what they took from it in life.

The story does not, however, follow this path; it, like so many stories, has its demon. In this case the demon quite probably lives in Liverpool, and he shows his destructive nature in buying Falkland sheep, and having them shipped to Britain for the population there to eat their flesh and spin their wool.

This is not destruction by commission, but by omission.

The effects are described by Sir John Orr:

> 'The process of the depletion and deterioration of the soil, which results in a decreased rate of growth and production, and in extreme cases by the appearance of disease, is proceeding on all pastures from which the milk, carcasses and other animal products are taken off, without a corresponding replacement being made.'

In Britain itself, domestic animals are reared and then sent off to industrial areas without replacement;

> '...in our own country this process of depletion has been going on for many years, especially in hill pastures, and it is probable that the decrease in the value of hill pastures in certain areas (due to the increase in the diseases and mortality of sheep) is associated with the gradual impoverishment of the pasture and its soil.'

The sheep want to get their full share in the life cycle, but they cannot. The minerals are not there; they left the life cycle when the sheep's ancestors were deported to industrial areas. Precisely the same has happened in the Falkland Islands, and for the same reason: that those in power value money more than life.

In certain grazing areas, however, there is a form of return; the pastoralists buy from external sources the minerals which the soil has lost, and these imported minerals are added to the depleted soil. It gives some temporary relief to the soil and this is perhaps being done in the Falklands; I have no information on the subject. It is possible that phosphates from the South Sea island of Nauru are being applied to the soil of the Falklands. Such a process would occur, and *does* occur in a great number of places, not through respect for the rule of return, but as an almost absolute necessity because the rule has not been followed. Were phosphates applied, there would be no need for any reconstruction of the soil of the Falkland Islands.

The herds kept by the British in the Falkland Islands constitute the biggest animal feeders upon the grasses. The grass draws its minerals from the soil, and these minerals pass into the bodies of the cattle. When those bodies, either alive or as carcasses, are taken out of the islands, then whatever minerals of the soil that they contain are taken out of the islands forever.

They arrive in Britain and there these minerals enter into the bodies of the British when they eat the meat of the animals. The British are, in effect, eating up the soil of the Falkland Islands.

To fulfil the rule of return, when ships take the animals to Liverpool or London, ships should be taking back their equivalent in soil nutrients to the Falkland Islands. This is no more impossible than it is for iron to float on the water in the form of the ships in which the food travels – *it is only a question of values and intention.* If the condition of the soil was the top priority, this would be done. If health were the top priority, it would be done. If the *economy* of health and quality were the top priority, it would be done.

But it is *not* done, because a small number of people profit from the ill health of the majority, and their business methods are inseparably involved in the breaking up of life cycles, and the consequent ill health and its social complications.

The merchant buys the products of the islands, but he gives back only something abstract – nothing more than his money.

The owners of the Falkland products exchange their goods for money, and they also pay no attention to the rule of return. They may be forced to buy some sort of manure for the land, and they do this under the pressure of the land's depreciation. As long as they can sell the natural wealth of the Falklands to Britain for money, they do so without any bad conscience as to the effect of their actions on the quality of life.

It is, in short, an astonishing anomaly that those who ship the byproducts of life to Britain are destroying the quality of life in the Falklands. Ultimately, they become as dangerous to the pastureland as the nomadic pastoralists of the last chapter were – in fact, being more powerful in means than the nomads, they are *more* dangerous.

What would be the result if the rule of return were followed?

This can be answered by consideration of a situation very different to the Falklands; an even smaller microcosm than the islands – a single dairy farm. The example I will now present suits our purpose well. It was a deliberate reconstruction, and one based on the rule of return.

*

The farm of 165 acres is situated within a hundred miles of the southern shore of the Baltic Sea, in a land of heath and pinewoods and an unfavourable climate of heavy wind and low rainfall. Its story starts with failure.

In spite of importing both sound cattle and feed at considerable expense, the farm fell, in the way that sick farms and industries often do, into the hands of a bank. The bank, of course, failed to make it flourish, and eventually it came into the hands of a farmer who believed that to be a sound and healthy farm, it must be a self-contained unit within the surrounding countryside.

The animals had to get their health from the soil on which they lived and not from outside, and they had to give back what they took from it. He proposed, as it were, to put a circle around the farm, and that circle was to be in effect a magic circle, protecting the inherent strength of the soil. A self-contained world of plants, animals and insects was to be created, and the balance that nature produces was to be given the chance to create wholeness and health.

Our farmer believed that only when the cattle can bind their whole nature with the soil that nourishes them can they and the soil together reach their full strength. Everything was planned as a whole, but for the purposes of our narrative the parts will be taken separately.

First comes the soil. Its chief needs were water, protection against wind, and nutrients.

Water was supplied by the annual rainfall, which was some thirteen to seventeen inches. More rain could not be got, but vegetation helped to prevent evaporation, as could be seen in the woods and heaths which were the natural cover of the land in the area. So the establishment of trees and hedges, with their double role of protecting the soil with their above-ground growth and connecting soil and subsoil through their root systems was implemented.

The trees provided a home to birds, and the hedges to flowers, hedgehogs, lizards, hens, hares and a variety of insects, all of which were neither encouraged nor suppressed, but freely allowed in the knowledge that each form of life seeks its sustenance from and gives its quota to the whole, and that *the resulting natural balance is the basis of health.*

Then came the plants cultivated for food; a smaller part for the humans and the larger part for the domestic animals.

The plants for the animals were chosen so that food from the soil was available all the year round. Where special protection for finer foodstuffs was necessary, a terraced garden system was devised. When the edible part of these plants had been used, the rest was rotted into manure by composting with the dung of the animals, and so returned to the soil. Nothing went off the farm except milk, and the occasional sale in later years of young animals, when their vigour and health had become widely known in the area.

We can now form a picture of this farm. It had become an aboriginal farm, in the sense of being true to its location. Added to this was the skill and knowledge that had deliberately reintroduced the methods of nature found in uncultivated areas, so as to regain nature's health and strength. The results of their efforts prove that it is never vain to ask nature to restore or sustain health.

The results could be termed miraculous, but the real miracle is not that nature is able to create and sustain such health, but that so few westerners understand and use this knowledge.

*

Let us now look at the conditions before and after the miracle.

When the farm was taken over by the new owner who was determined to enlist nature's help, the cattle were suffering from contagious abortion, and were strongly tuberculous. A number of animals had to be destroyed, and it was even debated whether it would not be better to destroy the whole herd to get rid of tuberculosis, contagious abortion and other diseases once and for all, and buy a new herd.

But the faith that works miracles was present. The herd acquired a new health. The new methods required understanding, hard work and sacrifice – but the result was that on this new self-contained farm, the soil soon revealed its surprising gifts.

The young plants were soon healthy and reproduced abundantly. The seeds that they bore were seeds which the sandy soil welcomed, as they were part of the same cycle as the soil itself. The plants looked strong and healthy, and the fodder straw preserved its beautiful golden colour. The plants grown for human use yielded foods rich in taste.

The development of the animals was the same. They became healthy and fertile, instead of sick and infertile. The cows gave more abundant milk, and the calves born on the farm grew to have double the output of the previous crop of milkers. The milk itself was rich in taste and acquired a special market amongst invalids, who enjoyed its taste and found its nourishing qualities beneficial.

At birth the calves born on the revitalised farm at once sprang to their feet and showed a lively temperament. They grew strong limbs and glossy coats.

Surprisingly, they contradicted the dictum that soils poor in chalk produce poor bones; these calves built notably strong bones despite living on chalk-poor soil. The fact, which in itself seems to be miraculous – that sunlight on the skin of beasts assists powerfully in the use of chalk – was now able to operate to its full extent, so that such chalk as the soil had was utilised efficiently. Every particle of it fell into its right place in the cycle.

As a consequence of all this, owners of neighbouring farms came to inspect the calves, felt their strong limbs, admired their vivacity and delight in life, and readily bought them to increase the strength of their own herds.

*

So we have these two contrasting pictures; that of the Falkland Islands – where the cattle showed such marked deterioration and where they were difficult to rear – and that of a sick and broken down farm which was turned around, and became an outstanding example of animal health.

The relationship between health and the wholeness of the farm is undeniable. Health is a positive quality, and it cannot be obtained or maintained except by wholeness in the cycle of life.

Diets, vitamins, and protective foods are *not* wholes. They are only parts of one factor (that of human food) in the cycle of life. The claim that they can produce real health, which is a whole, will always lead – or rather *mislead* – humanity to further disappointment.

Health is currently being pursued down nutritional avenues, as well as the anti-microbic sanitation avenue. But when the whole is the aim, the fragmentations which are created by these avenues – the specific microbes, the antiseptics, the sera, the vaccines, the chemical remedies which now receive the admiration of virtually all, the vitamins, the minerals of food, the protective foods, the hormones – all become unessentials, being absorbed by the positive whole, in which even the microbic world loses its negative and dangerous character, and becomes positive and beneficial.

Negatives vanish and positives take their place. The world, as fashioned by humanity, undergoes an enormous simplification. The many diseases of both humans and the animals and vegetables they farm constitute an immense mass of negatives, the elimination of which alters the very foundation of life.

At present we are constantly surrounded by new discoveries, inventions, and confused ideas about health, not to mention a vast array of life-destroying technologies and practices, simply because the lessons such as figure in this chapter have been ignored by the vast majority, and are understood clearly by only a few.

We shall now review further examples of situations brought about by the alienation of humanity from the creative power of the soil.

8
Banks for the Soil

THE TRADERS OF ENGLAND take living matter in the form of cattle from the soil of the Falkland Islands, pay money for it to the farmers, but pay nothing at all back to the living soil itself. The reason for this is that there are no banks in Britain catering for the soil of the Empire, although there are plenty of them trading in the farmers' money.

What the soil needs as payment for its share in the production and feeding of the cattle is not – of course – money. It does not want *symbols* of reality; it needs *reality itself.*

Unless it has this reality, it becomes less and less able to perform its role in the partnership between it and the Falkland farmers. So it must have a payment in its own currency – that of minerals and nutrients, and not in money. Of course, the money with which the farmers are paid *could* be turned into nutrition for the soil by means of more trade transactions – but that seldom happens.

The farmers, for example, could with the help of their banks buy phosphates from Nauru and apply it to the Falkland soil to make up for the phosphates that were removed in the bodies of the exported cattle. But this transaction would not prevent the final loss of the phosphates of the Falkland soil. These phosphates would just go down the drain – they would be eaten in beef or mutton by British consumers, be passed as excreta into the drainage system, and eventually reach some part of

Britain's Atlantic girdle. So, from the world of men, these phosphates would be dispersed into the vast, dark waters of the sea.

With banks for the soil in Britain, however, the story would be very different. Not only the phosphates, but all the life substances of the Falkland soil would be collected and paid back to the islands, just as the farmers' banks collect the money due to the farmers and pay it back to them. The banks for the soil would, in brief, obey the rule of return. They would do for the Falklands what they would also do for all exporting countries, the products of whose living soils are imported into Britain. They would collect all the various imported soil substances after use, transform them into soil nutrients and return them to the exporting countries from which they originated. Thus the balance of life – which is ultimately far more important than the balance of trade – would be preserved.

As it is, in this current age of commerce, nothing at all is done; the benefits of trade are siphoned off and separated from the processes of life as a whole. Quite unconsciously, the traders become the enemies of life, and they actually destroy that upon which their own wealth depends. They impoverish the soils, and in the end will so degrade them that trade will come to an end. Even the soil of the huge cattle estates of Argentina, which send far more animal food than the little Falklands to Britain, is known to be deteriorating.

It is, then, where traders and other business men are most concentrated that the need for banks for the soil is most urgent; it is there that the waste of nutrients is most colossal; and it is also there that the knowledge of this waste is so meagre as to be almost non-existant.

Britain, as an importing country, removes large quantities of raw material from foreign soils for use in food, clothing and manufacture. Many other countries do likewise, and the result is a huge transfer of the elements of life from exporting countries without any return – or in other words, a slow bleeding away of their resources. The importing countries act as blood-sucking parasites, draining the life of the exporting countries.

With vast territories, the exporting countries will enjoy a long period of prosperity founded upon the native fertility of the soil, but the end result is inevitable; a loss of the wholeness of the life cycles, partial or complete spoiling of the land, erosion, flood, the formation of swamps, barren hills and desert, degenerate plant and animal life, human depopulation and poverty, disease, and all the other consequences of the loss of soil fertility.

As for reconstruction; this loss of the fertility of the soil due to life-destroying commercial policies and activities can only be corrected by *the establishment of banks for the soil*, as has been previously described. It is not a question only of whether life can be healthily carried on without them, but of *whether it can be carried on at all.*

In 1896, Professor Shaler of Harvard gave a very clear and ominous reply to this question:

> 'If mankind cannot design and enforce ways of dealing with the earth which will preserve the sources of life, we must look forward to a time – remote it may be, but clearly discernible – when our kind, having wasted its great inheritance, will fade from the earth because of the ruin it has accomplished.'

That is the prospect with which the neglect of the soil confronts the peoples of this era of 'progress'.

The Falkland Islands are very small and very distant. Their loss would make little difference to the modern world. They represent only an infinitesimal part of the waste that is occurring worldwide on a truly enormous scale.

*

Let us now consider a prime example of this misuse of resources – the waste of human sewage. Everyone knows that manure can be turned into food by the soil, and in nature is returned to the soil; yet humanity's waste of this potential soil nutrient is vast.

The dictionary definition of waste is 'resembling a desert', yet what is correctly called waste does not *resemble*, but is the *opposite* of the desert. A desert is *devoid* of life.

It is modern water-carriage sanitation that takes the elements essential to human life and removes them from the cycle of life, and then calls them 'waste'. A picture of this waste is given by Professor King in *Farmers of Forty Centuries*:

> 'On the basis of the data of Wolff, Kellner and Carpenter, or Hall, the people of the United States and Europe are pouring into the sea, lakes or rivers, and into the underground waters, from 5,794,300 to 12,000,000 pounds of nitrogen, 1,881,900 to 4,151,000 pounds of potassium, and 777,200 to 3,057,600 pounds of phosphorus per million of adult population annually. This waste is regarded as one of the greatest achievements of our civilization.'

The loss of such quantities of the three elements mentioned is but a partial measure of the total loss, into the sea and other waters, of the elements of the human life-cycle, the loss of which could be avoided by the collection of these elements and their return to the soil.

To supply, by contrast, an example of a more thoughtful approach, King quotes Dr Arthur Stanley, then Health Officer of the city of Shanghai, in his annual report of 1899:

> 'Regarding the bearing on the sanitation of Shanghai of the relationship between Eastern and Western hygiene, it may be said that if prolonged life span is indicative of sound

> sanitation, the Chinese are a race worthy of study. While the ultra-civilized Westerner constructs systems for burning garbage at a financial loss and expels sewage into the sea, the Chinese use both for manure. They waste nothing, because the sacred duty of agriculture is of prime importance to them.'

Banking for the soil, therefore, captures Dr Stanley's attention. He was no advocate of sanitation systems for Shanghai that involved destructive methods of garbage disposal, or the modern ideas regarding water carriage systems.

There are in Europe, however, towns which, like those of the Far East, bank in the interests of the soil. There are towns which have actually gone back to this banking approach, after trying out the water carriage system. The capital of Sweden is one, and its transfer back to use, rather than wastage, was just completed when the Second World War broke out. In German towns, banking for the soil was ordered by the government in 1937. This policy was not based on the life of the soil. It was implemented as a war measure, and was an application of knowledge of life processes in the service of war and destruction.

In Britain, the Ministry of Agriculture in 1923 published a leaflet (No. 398) advocating this banking, but again not for the obvious reasons of the rule of return. Motor cars, buses and lorries had greatly reduced the number of horses in the nation, and therefore also the amount of available stable manure.

Among the various substitutes for this loss, and one which had great potential financial value, declared Leaflet 398, was ashpit refuse. There was plenty of it, but it was very little used.

> 'Incineration of this refuse is costly and is sheer waste. More up-to-date town authorities are now making an effort to dispose of their refuse in a better and more useful way, and some are adding other wastes and crushing the whole for use as fertilizer.'

There follows an account of what some of these towns were doing: London, Glasgow, Dundee, Perth, Aberdeen, Rochdale, Warrington, Halifax and in particular Gateshead, where 80 per cent of the houses had 'mixed pail' or ash closets, and hence the home refuse contained a considerable proportion of human excreta. This was crushed with ordinary town and slaughterhouse refuse, and made into manure at the low price of two shillings and sixpence per ton, at which price it was eagerly bought by local farmers.

Tested in the field, it showed itself to be a valuable substitute for farmyard manure. Nightsoil in dried form was prepared and sold by the Rochdale, Warrington and other corporations – a method, which if generally adopted, said the leaflet, would solve the problem of the wastage of sewage and

> '...the shortage of organic manures on the farm would be greatly relieved; but we must expect these methods of conservancy to be superseded.'

50 years ago, even London was a town from which farmers could take nightsoil for their fields, but since then, the excessive convenience of the modern water carriage system has earned it the approval, not only of the urban population, but also of many of the country folk, so that now any other method of disposal is viewed with suspicion.

One must expect these new forms of conservancy to replace the older forms which recognize sewage and garbage as valuable latent forms of life. That, however, is unquestionably what they are; and a hygiene which destroys them and drives them out of the human life cycle has no real right to its name.

The waste substances show their capacity to support life when combined. In the making of manure from the various town wastes, the materials, when mixed together, cook themselves through fermentation. A heap of compost, for example, can get so hot that if an iron rod is thrust into it, when withdrawn it is too hot to grip and hold. The final result of this process is similar to the leaf mould that forms on the floor of a forest. The rotting of vegetable and animal matter in a forest is a clean process, and that done in a town with urban wastes can be as clean, and as free from flies and smell. It can – and should – be a replica of what occurs in the forest, except that the pace of the urban method is rapid and makes good, sweet humus in three months, whereas the making of humus in the forest is a much slower process.

In both cases there is evidence of active life. The heat is one evidence, and the growth of fungus is so active that it can be seen like smears of whitewash both on the floor of the forest and in the urban heap.

This waste announces, in an outstanding manner for something supposed to be dead and done for, that it is very much alive, and that it is just as much a part of the life cycle as when it pulsed in the hot blood of a Derby winner or a milk cow. It is also hot life, indeed, for during the greater part of its activity it is considerably hotter than an animal's blood. Then, when its heat and activity die down, it becomes aromatic and crumbling humus, and a starting point for the rich green growth of healthy vegetation.

Such is one result of banks for the soil. The shamefully misnamed 'waste' becomes beautiful, soft, crumbling humus, and the very substance of healthy life. It needs, perhaps, a poet to realize what beauty it contains. A poet can see in it the great, positive *'YES'* which is the unchangeable token of healthy life and of all that gives strength, grace, swiftness, endurance, cheerfulness, agility, elegance and beauty to mankind.

It is the universal parent of all that is excellent and positive in life.

9

The Economics of the Soil

THE LINK WHICH CONNECTS town and country in the use of urban waste is financial. One would be unable to find a municipality that turns its wastes into humus purely for the good of the soil – those that do so do it only because farmers will buy it.

By selling that which had once been waste, it is converted into cash, and this is a good result for the municipal ledgers. Leaflet No. 398 in the previous chapter, of course, had no choice but to accept the dominance of this viewpoint. The leaflet sought to persuade the farmers to make use of prepared waste, so that the great loss to the land of organic manure which had resulted from the diminution of stable manure could be mitigated, or even fully compensated for.

Although the costs (in particular those of transport) involved prevented farms distant from towns being able to get the manure, those close to towns were urged to use it. As things were, the leaflet stated that some 10,000,000 tons of ashpit refuse was produced annually in England and Wales, and that towns were spending £6,000,000 a year on collection and disposal. This was unquestionably waste.

The issue of money, therefore, overrode any other concerns. Twelve years after the publication of the leaflet of 1923, Sir George Stapledon stated that sixteen and a half million acres of England and Wales, or 43 per cent of the total of cultivable land, had fallen into 'a more or less neglected condition'. They were however, he said, 'capable of radical improvement'.

The Earl of Portsmouth, about the same time, summed up another aspect of the same question:

> 'It is a staggering commentary on our present attitude to health and agriculture that, excluding all accidents, all patent medicines and private medical cases, the bill for sickness in this country amounts to £276,000,000 a year, while the farmers receive for their gross output barely £250,000,000.'

To put these two pieces of information together; there was a great deal of wastage, both of waste products and of land itself. The waste of land was very great indeed. This was more surprising in that it occurred in a country threatened by war on a greater scale than that recently experienced in 1914–18; a war in which it was nearly brought to its knees by the lack of well-cultivated homeland, and in which the blockade of its enemies and their consequent shortage of food was a large factor in their collapse.

The need for good soil had been so emphasized by world events that it would have been inexplicable that it was not regarded as being of paramount national importance – were it not for the fact that, for a prolonged period, a cultural barrier had both practically and intellectually alienated the bulk of the population from the soil.

So a strange happened. In spite of the great danger of the neglect of soil being written large in letters of blood, the people were blind. They were also deaf, for they ignored the warnings of such authorities as Sir George Stapledon, the Earl of Portsmouth, and many other leaders of the countryside. As Thoreau said:

> 'It takes two to speak the truth. One to speak and one to hear.'

The people were indeed unable to hear. The problem was the dominance of money – of the cash nexus. As long as concern for money was paramount, the creative power of life and all that pertained to it was inextricably fettered. Only a rare individual could escape from the entanglement and see things in their proper proportion.

Such a person was the late Oswald Spengler, author of *The Decline of the West*, and his account is so clear that it must here be given in his own words. (Spengler's German is very difficult. My quotations are from the two-volume English translation of his work.) He begins his analysis at the time when civilization was purely agrarian. The life of the bulk of the population is purely that of the peasant on the open land; the town, with its fully urban character, has not yet come.

The most important institution in this world of villages, castles, palaces and monasteries is not a city, but the *market*, a meeting place defined by yeomen's interests. It also had religious and political meaning, but certainly cannot be said to have any special life of its own. The inhabitants, even though they might be artisans or traders, were still essentially peasants.

> 'That which separates out from a life in which everyone is alike producer and consumer are *goods*, and traffic in goods is the mark of all early commerce, whether the object be brought from the far distance or merely shifted about within the limits of the village, or even the farm.
>
> A 'good' is that which adheres, through its essence, to the life that has produced it or the life that uses it. A peasant drives his cow to market, a woman puts away her finery in the cupboard. We say a man is endowed with this world's 'goods'; the word *possession* takes us back right into the plant-like origin of property, into which this particular being – and no other – has grown, from the roots up.

Exchange in these periods is a process whereby goods pass from one circle of life into another. They are valued with reference to life, according to a sliding scale of perceived relation at the moment. There is neither a conception of value nor a kind or amount of goods that constitutes a general measure – for gold and coins are goods too, but their rarity and indestructibility cause them to be highly prized.

Into the rhythm and course of this barter the dealer comes only to intervene. In the market the economics of acquisition and creativity encounter one another, but even at places where fleets and caravans unload, trade only appears as an *organ* of countryside traffic. It is the 'eternal' form of economy, and it is even today seen in the ancient figure of the pedlar of the country districts remote from towns, in the out-of-the-way suburban lanes where small barter-circles form naturally, and in the private economy of savants, officials, and in general everyone not actively part of the daily economic life of the large cities.

With the development of larger urban centres, a different kind of life awakens. As soon as the market has become the town, it is no longer a question of being a mere centre for a stream of goods traversing a purely peasant landscape, but of a second world within walls, for which production 'out there' is nothing but object and means, and out of which another stream begins to circle.

The decisive point is this; *the new urban dweller is not a producer in the earthly sense.* He does not have the connection with the soil, nor with the goods that pass through his hands. He does not live with these, but sees them as something external, and regards them primarily

> in relation to the upkeep of his own life. Goods become nothing but wares, and exchange becomes turnover, *and in place of thinking in goods, we now have thinking in money.*
>
> With this, a set of limiting definitions is abstracted from the visible objects of economics, just as mathematical thought abstracts something from the mechanistically conceived environment. Abstract money corresponds exactly to abstract number. Both are entirely inorganic.
>
> The economic picture is reduced to *quantities*, whereas the important point about 'goods' had previously been their *quality*. For the early-period peasant 'his' cow is, first of all, just what it is – a unit or being, and only secondarily is it an object of exchange. In the economic outlook of the new townsman, however, the only thing that exists is an abstract monetary value, which at the moment happens to be in the shape of a cow, and which can always be transferred into that of, say, a banknote.
>
> In the same way, most modern engineers will see in a waterfall not a unique natural spectacle, but merely a calculable quantum of unexploited energy.
>
> It is a fundamental error of all modern financial theories that they start from the value token or even the material of the payment token, instead of from the form of economic thought. In reality money, like number or law, *is a category of thought.*'

The above is clarity combined with profound insight. The original model of the agrarian world, in which production is rooted in the soil, gives the products a reality because of the quality or life that is embodied in them. They become man's possessions, something near him, and are valued with reference to life.

But within the soul of the town a quite different kind of world view arises, one in which something intervenes between 'goods' and man. The result, in its essence, is the change from creative goods-based thinking to abstracted money-based thinking. The difference is expressed here by Spengler:

> '...goods become 'wares' (things of the warehouse, not of the personal home), 'exchange turnover' (i.e. not a mere interchange of goods for other goods), and in place of thinking in goods we have thinking in money ... in reality, money, like number and law, is a category of thought.'

Let us look closely at this difference, particularly in its relation to those 'goods' which are most closely related to life and without which life could not exist – the food products of the soil: seeds, roots, leaves and fruits.

Early humanity made the observation that when seeds were put in the ground, plants grew up which produced a far greater number of seeds than were put into the ground. The early farmers did not worry about whether or not the production of a great number of seeds from a few plants revealed a rather gloomy (and even brutal) design on the part of nature, intended to make only a few seeds successful in becoming plants, and so proving their superiority to the rest as being the fittest and most able to survive in the struggle for existence. They did not regard the few and the many as being due to a rather snobbish, if divine, order of precedence.

They were more simply concerned with the fact that these extra seeds, tubers, fruits and foliage provided them with food, and so they saw themselves as being involved in the ongoing process of creation. All they understood was what they saw the generous outpouring of abundance, in response to their efforts, and thanks to a mystical power which in its working was beyond them, but in its revelation to them aroused their awe, their reverence, and their gratitude. So they served nature to the best of their ability, drew their share from the cornucopia of abundance, and humbly thanked their gods for this revelation of divine love and magnanimity.

This abundance, created as the result of their labours, enabled families to supply food not only for themselves, but also for others. A certain part of their produce was, therefore, set aside for the non-working members of the families, for craftsmen who gave them goods in return for food, and for the government in the form of taxes. And this is important: *they did not pay their government for its services in money, but in produce.*

Now the great significance of this payment of tax in kind is that it is, of course, completely related to the basis of human life; the soil.

The soil yields grain, of which a portion, say a fourth, is set aside for the government. Taxation, therefore, depends on the stability of the soil, and nothing is so stable in human life as well-cultivated soil. There is, when the products of agriculture are many and varied, a steady return from the soil, if the agricultural techniques employed are not wasteful. There *must* be some such steady relationship between the soil and humanity if human life is to continue without violent fluctuations. There are, of course, good seasons and bad seasons. There are times of drought, flood, and disaster, but a settled and capable agriculture nevertheless supports prolonged national life and wellbeing.

Payment of taxes in kind is a payment in terms of that which is primary to national life. It is factual and real in a whole national sense. It is true to the earth and sufficient, and has no foreign, extraneous and unlimited character, such as life dependent on conquests, on the destruction or annexation of weaker nations or, through the agency of money, on the well- or overfed condition of the few and the underfed condition, or malnutrition, of the many. In short, it liberates the soil and keeps it free from money – the one real and essential freedom necessary for a whole national life.

Under the old system, the peasants paid the king for national protection. That was the service he rendered to them and for which they returned reciprocal service. That is the doctrine to be found in the classics of the past, such as the *Smriti*, or law-books of ancient India. The land of the country was not the king's property, but the common property of all who worked that land and enjoyed the fruits of their labour, as Professor Dvijadas Datta states in *Peasant-Proprietorship in India* (1924).

Taxes once existed to protect both the living land and the land of the living, and not, as they have now become under the priority of money, that which allows things so anomalous as, for example, the payment of interest on money lent by the privileged class for their long past wars. The peasants did not pay the king to protect the land against enemies whose dead bodies had long been dissolved into the soil; they paid for the protection of the land on which they were living, and by which their nation was supported.

No one can juggle with the soil as thoroughly as humanity, in its greed, has learnt to juggle with money. The soil is reality. It has its own character: it is more powerful than humanity, for it holds the key to the infinite mystery of the power to turn death into life.

Money, on the other hand, is purely a human invention, and we can fashion from it whatever we like, from the ponderous blocks of iron of the honest Lycurgus to the meaningless book entries of modern bankers or manufacturers of credit. It can take every form of transubstantiation that we choose to put upon it. It permeates everything that humanity dominates. It is only on the land that humans will ever be able to be free of money. It is only there that we will be able to see clearly what life really is.

And life is something that starts from the health of the soil in a way that, if it is to be successful, the principle of life must direct. Soil, in conservative and whole life, should direct and rule money – money should not rule the soil. Soil is the primary thing, and in reconstruction its needs must be provided for, without money having priority over it.

Money should act rather as a balance, as a servant to the soil; so it acted at least amongst Indian and other peasantries. That is why it was denoted by metal, and why it was recognized as a possession because, being metal, it had durability just as the land had durability.

Money could act as a substitute for the land. When there was scarcity in local produce, money made stored food and second-class food available, for example by assisting poor land to be cultivated. When famine threatened, then the silver bangles of Indian women were taken to the *sowcar* and weighed by him and turned into an equivalent weight of silver coins, so that coiniage became more plentiful in times of distress.

This is the *exact opposite* of urban banking. When distress threatens, bankers call in their loans. As distress increases, the amount of money in circulation becomes less, not more – more distress means *less* local money, rather than distress leading to *more* local money. In very great distress, according to the sages, it was right for the king not only to forgo the taxes in kind, but to give money – not loan it – in order to lighten the distress, by enabling people to buy food and assistance from outside their locality.

The proper economics of the soil do not exist under the dominion of money. If the soil is lined up with other productive agents of saleable goods, its intrinsic character and value vanishes. It is essentially different to goods manufactured for sale, for it is as much the property of life as is the air. Neither soil nor air have rightful market value because they are necessary means of life.

There is no market value (yet) for air, nor should there should be one for soil. City air, burdened with petrol fumes and other pollutants, is not just bad economics, but bad life. Soil that is burdened with money is not just bad economics, but bad life also. That is why the right human partnership with the soil must be an essential part of human life if it is to endure and prosper.

With the right conservation and service, the soil responds with something that is as reliable and stable as the human actions which, through the continuity of family service, provide this protection. It responds with its gifts with a regularity which is entirely different to the violent fluctuations in national and personal life which have occurred under the influence of the precious metals, and owing to which the most profound effects in modern civilization have followed upon the discovery of Potosi silver, Californian gold and improved chemical processes for extraction.

Nothing, one feels, could be more fantastic and unattainable than to try to stabilize human life – and it *must* be stabilized if catastrophe is to be avoided – while measures of such inconstancy are permitted to dominate.

*

Let us now, then, in the midst of our modern inconsistancies and the great catastrophes which threaten us, review this great virtue of *constancy*, in terms of the creative power of the soil. Here we turn to Professor King's *Farmers of Forty Centuries*.

His introduction describes the contrasting pictures of the thorough and profound relation of humanity to the soil in China and the resulting conditions of social constancy, compared with that of the undeveloped relation to the soil of Western culture.

He describes the example of the meticulous care with which water is preserved and used in China:

> 'To anyone who studies the agricultural methods of the Far East in the field, it is evident that these people, centuries ago, came to appreciate the value of water in crop production as no other nations have. They have adapted conditions to crops and crops to conditions to such a pitch that in the case of rice, they have produced a cereal which permits the most intense fertilization and at the same time ensures the maximum yields in cases of both drought and flood.
>
> With the practice of Western nations in all humid climates, no matter how completely and highly we fertilize, in more years than not, yields are reduced wherever there is a deficiency or excess of water.'

He went on to summarize the extent of the canal systems of China, a conservative estimate of which would place the miles of canals at 200,000. China has as many acres in rice each year as the United States has in wheat, yet that land does not bear rice alone, but produces at least one, and sometimes two, other crops each year.

When and where water is not available for irrigation, the people cultivate quick-maturing, drought-resisting millets as staple food crops, and for them the water is preserved by:

> '...almost universal planting in hills or drills, so making possible the utilization of earth mulches in conserving soil moisture.'

Thus

> '...these people have, with rare wisdom, combined both irrigation and dry farming methods to an extent and with an intensity far beyond anything that our people have ever dreamed of, in order that they might maintain their dense populations.'

The canals, moreover, render not only water, but a refreshment of soil itself comparable to that of the overflow of the Nile or of the warping of the Isle of Axholme.

> 'In China, large quantities of canal mud are applied to the fields, sometimes at the rate of 70 or more tons per acre.'

And where this mud is not available, the canals still yet refresh the soil in a manner again rivalling the autochthonous renewal of Egypt.

> 'So, too, where there are no canals, both soil and subsoil are carried into villages and there they are, through the application of great labour, composted with organic refuse, then dried and pulverized, and finally carried back to the fields, to be used as home-made fertilizers.'

Finally, on page 241, he states that:

> 'China, Korea and Japan long ago struck the keynote of permanent agriculture... In selecting rice as their staple crop; in developing and maintaining their systems of combined irrigation and drainage, notwithstanding that they have a large summer rainfall; in their systems of multiple cropping; in their extensive and persistent use of legumes; in their rotations of green manure to maintain the humus of their soils and for composting; and in the almost religious fidelity with which they have returned

> to their fields every form of waste which can replace the nutrients removed by the crops.
>
> These nations have demonstrated a grasp of essentials and of fundamental principles which should cause Western nations to pause and reflect.'

It is clear that in these works and actions of the Chinese, all the factors which promote the fertility of the soil are brought together so as to ensure and preserve its highest creative power. This is done at the expense of great labour, so as to conform to the true character of the economics of the soil. By this labour a constancy of return from the soil can be assured; a constancy which has no parallel in the dominant financial system of our time, a constancy which depends upon the fact that if all the factors of fertility in a locality are brought into the service of cultivation, the results will reach a high degree.

The concept of the dominance of money is, on the other hand, foreign to the soil. When money is lent, it expects to get not just itself but *more* than itself in return. Even disregarding the speculative hopes of capital improvement, money lent expects an addition of itself called *interest*.

On the other hand, in good agriculture, fertility is fully used in producing a crop. It is not and cannot be called upon to create an extra quantity of itself so as to produce an extra crop or 'interest'. Only something parasitical could expect to receive interest, and that does not occur in whole farming. In farming dominated by money, however, parasitism is as abundant as debt; like breeds like. If one reads a book on modern farming one cannot help but be struck by the number of parasites that take their share in it. There are warble flies, scabs, lice, fleas, maggot flies, bollworms, eelworms, wireworms, fruit flies, fungi, leaf roll, blackscab, blight, mosaic, rust, bunt, smut, leaf stripe, black leg and so on. The more complex that scientific farming becomes, says D. H. Robinson,

> '...the greater is the spread of complaints which formerly were unknown or of little importance.'

There is a clear difference between a farm devoted to the preservation of high fertility and one devoted to the production of money crops under the dominance of money, credit and debt. Once a farm is involved in the credit debt system, and once this credit debt is looked upon as a first need or being of primary importance, agriculture becomes inextricably involved in a huge system, complete with its owners, its managers, and its local, national and international debts.

These debts affect everyone within the system. Modern humanity, in facing the problems and complexities of life, finds itself loaded and hampered by the dead and crushing weight of debt. The size and the pace of growth of these debts are so extreme that there is no hope of their being balanced by the creative power of life.

The only response to this debt is to use up without replacement the stored fertility of the past, but even this fails eventually. It does not abolish, but ultimately *extends* both debts and debtors on the land.

The whole situation is so utterly unbalanced that only people with minds totally divorced from the reality of creative life could possibly acquiesce to the hypotheses and creeds which have arisen to fortify it and to make it appear rational and sane – hypotheses which have raised the speculator and the millionaire to the status of darlings of nature; her selections in the survival of the fittest!

The stark fact that appears now, and which wrote itself across the decline of the Roman Empire, is that *debt and taxation increase as the soil declines*. One is a counterpart of the other. The huge, unpayable debts are the measures of the death of reality; step by step they are matched by the loss of soil fertility.

In coming chapters, we shall see how remarkably the dominance of money in the present era is matched by the increasing ill-health of the soil. The modern financial system and its vast debts - personal, local, national and international - are firmly on the side of death, and work against the creative power of life.

Nature, it must be remembered, has no inherent interest in maintaining a highly organized form of life such as we humans. If we take a harmonious place in the cycles of life, humanity will continue; if not, we will be replaced by some other form of organic life, just as surely as bracken replaces grass.

Survival is not a matter of struggling to be fittest, it is not a matter of modern boasts of the conquest and exploitation of nature.

It is a matter of reverence.

10

The English Peasant and Agricultural Labourer

THE ENGLISH PEASANT first appears in 'Engleland', as an individual with a strong bent for independence. Engleland was the southern part of the thumb of land that projects itself between the North and the Baltic Seas, the northern part being the land of the Jutes, or Jutland. The people of Engleland, writes John Richard Green, in his *Short History of the English People*,

> 'seem to have been merely an outlying fragment of what was the Engle or English folk, the bulk of whom lay probably along the middle Elbe and on the Weser.'

He adds that they were allied to peoples occupying a wide tract reaching to the Rhine, and collectively known as Saxons. Green does not, however, speak of the fascinating theory of Henri de Tourville, who gives the name of 'particularist' to these Nordic peoples, because they were people of the small (or particularist) families consisting of husband, wife, and children as opposed to large extended families of fathers, their sons and grandsons, and their wives and children.

Henri de Tourville, in his *Histoire de la Formation Particulariste*, believes that this small family came into being when Teutonic or Nordic people reached the plains of Sweden and in their search for a safe home passed on over the mountains and settled along the fiords of Norway.

Anyone who has voyaged up these fiords has seen the

patches of bright green cultivation that are set between the precipitous mountains and the sea. They are like gems of emerald. The traveller will also have been struck by the smallness of most of them; nevertheless, what is grown on them, and the fish caught in the fiords, form an adequate food supply for these isolated families. These families were small because of the limitated supply of vegetable food. When the families of a fiord grew too large, the younger members gathered together, stocked a few ships and set forth, seeking land in fiords farther south, in the projecting thumb of Denmark, in the northwestern river-lands of Germany, and finally in the island of Britain.

In the new settlements, their love of independence assured the persistence of the small family system. However this system actually arose, it has been of great significance in the world's history. It is the oddity, as opposed to the customary large or joint family; it is independent individuality as opposed to dependence on joint opinion; and a very strong oddity it has proved to be.

However rude and rough these early Engles may have been, there are few Englishmen now should will not be thrilled when they read how Tacitus, coming from the great city of Rome, was struck by the jealous independence of each farmer and his family in their settlements. He wrote:

> 'They live apart, each by himself, as woodside,
> plain or fresh spring attracts him.'

They could not, however, be totally independent. Dangers from other peoples sometimes threatened them, and they then joined together, chose a chief and took to arms. They were fierce fighters and when they arrived in Britain, they drove the Britons westwards or slew them, and took their land, until once more they were independent farmers at peace. They were the forerunners of similar settlers in America, Australia and New Zealand.

But before the coming of the Norman conquerors, these farmers, says Green, had already lost most of their peace and independence. They had so many wars that warrior-kings and their military subordinates had become a standing feature of their society. For greater protection against invaders, similar to themselves in race, they had to submit to larger associations, and eventually were forced into one kingdom.

They lost their spontaneity of action and had, as a condition of existence, to attach themselves to a lord, or *thegn,* of the King's party. Green wrote:

> 'The ravages of the long insecurity of the Danish wars drove the free farmer to seek protection from the *thegn*. His freehold was surrendered to be received back as a fief, laden with service to its lord. Gradually the 'lordless man' became a virtual outlaw in the realm. The free churl sank into the villein, and changed from the freeholder who knew no superior but God and the law, to the tenant bound to do service to his lord, to follow him in the field, to look to his court for justice, and render days of service in his demesne.'

The coming of the Conqueror, William of Normandy, confirmed the subordinate position of the English farmers, by giving them foreign conquerors as their lords. Green wrote that the increase in the authority of the aristocracy

> '...was quickened by the conquest. The desperate and universal resistance of his English subjects forced William to hold by the sword what the sword had won, and an army strong enough to crush at any moment a national revolt was necessary for the preservation of his throne. Such an army could only be maintained by a vast confiscation of the soil. The failure of the English risings cleared the way for its

> establishment; the greater part of the higher nobility fell in battle or fled into exile, while the lower *thegns* either forfeited the whole of their lands, or redeemed a portion of them through the surrender of the rest.'

Land became the property of the King, who rewarded his followers and bound their interests to his through gifts of land. The Norman aristocracy received many estates, scattered so that they could not constitute a dangerously strong local power, but even the meanest Norman could rise to wealth and power in the new dominion.

So William initiated land as the private property of an aristocratic caste of landowners, and the peasants became serfs.

England was, in terms of population, a very small country at that time. It held two million at the time of the Conqueror, and two and a half million at the time of Edward III. The total area of cultivated soil was small, the greater part of the land being forest, and therefore possessing undisturbed its primal vegetative cover.

The farming was backward, as the slow population growth reveals, and, compared to that of more enterprising countries on the European continent, it remained backward for many centuries. Nevertheless, it maintained a life cycle which, though of low grade, preserved within itself a certain stability and was free from pronounced or destructive waste.

When a balance between the English and their Norman conquerors was finally achieved, a new association based on the soil (the features of which the reader should now be acquainted with) came into being.

Farming was carried out on large estates. These estates were called manors, and the heads of the estates were the lords of the manor. Under these lords the people worked, with various levels of right to the land, through which one and all got their food and home directly from the land.

The country as a whole operated under a 'natural economy', not a 'money economy', and such commonplace features of today, such as capital, labour, competition, and employee had no meaning.

The family or associative method was everywhere. A man might employ labour, but he worked beside those he employed, and he ate the same food as they did. The manor was, indeed, like a large family. It was a self-contained community, and the land itself was the father and mother of the community. The lord of the manor represented a personal government, but he was not able to do with the land whatever he wished. His position was that of chief functionary, and not that of slave owner, as in post-Punic Italy.

The land was worked on a common plan. There were no separate fields, but one large open space, marked off into strips by balks. The lord of the manor would often have his strips amongst those of the villagers. In such cases the community was a true community, in which the land was alike to all, but in other cases the personal land of the lord of the manor was not amongst, but separated from, that of the villagers.

In addition to farming by the manor system, the most educated section of the population – the monks of the Church – contributed to the farming culture of the nation the benefits of their devotion, learning, and art.

William Cobbett has given an account of the special character and quality of the monasteries and their meaning in an agricultural civilization, in *The History of the Protestant Reformation*, written over a hundred years ago. He wrote:

> 'Nor must we by any means overlook the effects of these institutions on the mere face

> of the country. The man who is insensible to any feeling of pride in the noble edifices of his country must be low and mean of soul.
>
> The monastics built, as well as wrote, for posterity. The never-dying nature of their institutions ignored, in all their undertakings, any concern as to time and age. Whether they built or planted, they set the generous example of providing for the pleasure, the honour, the wealth and the greatness of generations upon generations unborn.
>
> They executed everything in the very best manner: their gardens, fishponds, farms, were as near perfection as they could make them; in the whole of their economy they set an example that made the country beautiful, made it an object of pride with the people, and made the nation truly and permanently great. Go into any county and survey, even to this day, the ruins of its abbeys and priories, and then ask yourself: "What have we given in exchange for these?"'

To their practical farming, the monks brought the help of the classic writers of Rome – Cato, Varro, Columella and others – whose works in Latin they were able to read. They were *cultured* farmers, to whom the spiritual side of creation appealed with special significance.

It was they who instituted improvements, and preserved a high standard in medieval farming. It was they who harboured the endeavour to do well, without which the work of the mass of humanity tends to decline. It was they who built roads and bridges, and opened their monasteries as places of temporary rest and hospitality to all travellers, rich or poor; they who drained marshes, reclaimed wastes, and improved livestock. It was they who sustained a soil-based civilization by giving it the vision of religion, the art of the temple, and the culture of studentship. They also defended, as far as they could, the independence of the peasants, and supported them in their efforts to rise out of serfdom.

The lords of the manor were the worldly leaders of the people. They supervised and directed the division of the land, saw to the upkeep of cottages and buildings, presided over schooling and apprenticeship, arranged marriages, punished slovenly work, dealt with quarrels and crimes, checked short weights and the adulteration of grain and beer, arranged for the exchange of goods, and directed the relations of the villagers with the outer world that began on the farther side of the forest that bounded the manor.

We now come to the introduction of the 'money economy'.

At the time when the manor system flourished best, the lords of the manor were the paternal chiefs of the villagers, but they also had a number of rights that came to them through conquest and were, in fact, derived from the Norman Conquest. It was these rights that made their precedence in the village something different from that of the village assembly, which had previously been the common form of village rule and which constituted the true freedom and independence of the partners of the soil.

The lords of the manor possessed the right to exact a varying amount of enforced work from the villagers; they exacted fees for the services of the manorial court; they had the right to sell timber from the estate, to permit strangers to take up land, to mill and even bake the people's bread; and, their class being the lawmakers of the country, they were able to pass such laws as the *Statute of Merton* in 1236, which gave them the right to enclose certain lands of the villagers for their own use.

In brief, they were undisputed masters; they prolonged the Conquest indefinitely, and thereby prevented the villagers of England from possessing with complete freedom the land they cultivated.

There was one other privilege possessed by the lords of the manor which was in direct contradiction to the freedom of those who worked the soil. It was this; they had the right to fold, not only their own cattle, but also those of the villagers, on their land. They became the 'manurial', as well as the manorial, lords of the estates. Everyone in the village, of course, knew that when they took the manure, their lords were robbing them of food.

The lords of the manor, seen from the viewpoint of the soil, became thieves in the midst of the village. They were stealing the villagers' cow dung long before they openly became robbers and pillagers under Henry VIII. Through their theft or privilege – whichever it be called – their land received greater (and that of the villagers received less) fertility.

As a result of this change in the soil, there followed a change in the population; a difference in quality arose. The rich, fed by fertile soil, enjoyed better health. The health and quality of the people generally, however, was degraded. Poverty and wealth became not only a variable measured by money, but also a visible physical condition.

There is nothing that should be made more clear than this: the first separation leading to the divided classes of employers and employed, of rich and poor, with the poor dependent not on the soil but on the rich, was – as unlikely as it may sound – *the separation of farm dung.*

It was a personal sequestration of life elements. It was not a crime in English law, but in terms of the soil it was a lethal crime that eventually led to disasters for the robbed.

Immediately, the life cycle of the lord's demesne was improved, and that of the peasants' land was diminished. Lord Ernle, in *English Farming, Past and Present*, wrote:

> 'On land which was inadequately manured, and on which neither turnips nor clovers were known till centuries later, there was no middle course between the exhaustion of continuous cropping, and the cure of resting the soil.'

Much of the land had to lie fallow, unused and uncultivated until it was able to recover the strength which the lords of the manor had taken from it.

The aristocracy needed the extra wealth which this sequestration of life elements brought them. The crime was forced upon them by their luxury and expenses as courtiers, and as their roles as warriors in the Crusades and French wars. They became, consequently, exactors, not protectors, of the soil, and they displaced the old natural economy of the manor with the new money economy.

The more enterprising and frugal villeins of the manor, supported by the Church, saw in this need of their lords the opportunity to satisfy their cravings for independence. With the surplus they achieved by their ability, they won their freedom from service to their lords and they became tenants through the payment of rent. They took over land, too, from the least efficient of the manor's farmers and worked it using the previous owners as labourers.

Thus, during the slow break up of the manor system owing to the introduction of the new money economy, the people of the manor came to be divided into four classes; the first was the lords and their families and personal dependants; the second the tenant farmers; the third the villeins, who did not become tenants; and the fourth those who failed to support themselves upon the land that had been allotted to them, and who now worked for their more successful brethren for a wage, paid in kind or in money.

This fourth class are often spoken of as the class of free labourers, because they were to some extent free to sell their labour. Their freedom was very limited though, because their poverty compelled them to use it as labour uses its freedom today; in binding itself to this or that master.

They lost their right to the land and to the stock which had been their capital. Their value was relative to their abundance or their shortage. Only when there was a great shortage of labour, such as that which followed for many scores of years the destructive Black Death of the middle of the fourteenth century, did their wages exceed the cost of their necessities. Thorold Rogers called the fifteenth century the golden age of the English labourers or farm workers, as measured by the relation of their wages to the prices of their necessities.

The freedom that these relatively high wages brought was eventually defeated by the continuous decline of the soil in the early Tudor period. Lord Ernle wrote:

> 'Land had depreciated in value; rents had declined; farming had deteriorated; useful practices had discontinued; cattle were dwindling in size and weight; the common pastures had become infected with 'murrain'; the arable area of open fields had grown less productive, and without manure its fertility could not be restored.'

Desperate measures were required to save the land, and these were undertaken were those empowered by the ascendant money economy.

In Roman Italy, after the Punic Wars, the deterioration in fertility of the soil led to the substitution of family-owned farming with large estates, the *latifundia*, and large landowners. In Tudor England the same substitution of *latifundia* for small family farming also took place. In post-Punic Italy, acquisitive men seized the lands of weakened farmers with complete disregard of the law.

Mommsen tells us:

> 'The whole system was pervaded by the utterly unscrupulous spirit that is characteristic of the power of capital ... Roman capital was gradually absorbing the intermediate and small landed estates in Italy as well as in the provinces, as the sun absorbs drops of rain.'

In Italy, the large number of slaves acquired by Rome's conquests hastened the process, for it was easy for large landowners to break right away from their own fellow countrymen, and, leaving them to their fate, to engage foreign slaves to work the Italian soil. In England, the process of the eviction of peasant family farming was not completed until the advent of the industrial era.

In both cases, as on similar occasions elsewhere in history, the social change took on the nature of a conquest. A class of acquisitive men who had obtained money through means other than direct agriculture acted as conquerors. They overthrew the role that the peasants' customary rights to the soil had played as the basis of the State, and made land a commodity to be purchased by the richest bidder.

In Italy these acquisitive men were the Equites or Knights, who had acquired great wealth by acting as middlemen, and who were to form the chief part of the aristocracy of what would become the Roman empire.

In England the acquisitive men who overthrew the agricultural basis of the State and with it the Church and the monks became the new aristocracy of Tudor England.

In both cases, there were also statesmen and other leading thinkers who opposed the changes under which the independence and rights of the farmers and free labourers were to succumb. Among these were Wolsey, More, Latimer, and Queen Elizabeth and her Ministers amongst the English.

Nevertheless, in spite of all such efforts, the great living fact about soil remained. It was expressed by Ernle:

> 'Without manure its fertility could not be restored.'

Dung had to save the soil, and the quickest way to apply dung to the land was to enclose it with hedges, and breed and put upon the fields sheep and cattle. Fortunately, the acquisitive men were attracted to this method by the price that British wool fetched upon the Continent. It was this opportunity for more wealth that made them seize the land of the small farmers and the monasteries, and with the expenditure of their capital turn it into sheep farms. It was unquestionably good for the soil, but it entailed a brutal punishment to the small farmers and farm labourers whose only sin had been that they had submitted originally to the enclosing of the lord's demesne upon the manor, and the robbery of the dung of their animals. In this manner, a new aristocracy arose from the human relics of a system that had failed, and the brilliant later Tudor period of English history followed.

From that time onwards, the proletariat and poverty became familiar parts of English society.

No appreciation of the value of the small holdings was evident. There was no Prince Kropotkin to propose that, with the intensive farming of small holders, the British soil might support a hundred million inhabitants. Nothing was known of the achievements of the Chinese peasantry who were so skilled in the use of water, and who followed the rule of return with such meticulous care. Nothing was known of the agriculture of the fallen Arabic Empire. The Tudor world was deeply stirred by what Green calls the New Learning, but the New Learning did not bother itself with the humble giver of life, the soil.

For the full story of the English agricultural labourer,

the only authoritative history in English that I have been able to find is *A History of the English Agricultural Labourer*, by Dr W. Hasbach of the University of Kiel. It was first published in 1894, translated into English in 1908 and reprinted in 1920.

Where enclosure occurred, Hasbach says, a proletarian class appeared. English agriculture from the fifteenth century, when rich commercial men began to buy out owners living on their land, was sacrificed to the interests of industry.

He gives a full account of the second great period of enclosures, that of the eighteenth century. It was in the latter part of this century that the combined genius of the English and Scottish ushered in a new epoch, that of the machine. The power of these machines effected a revolution. Manufacturing towns grew up and multiplied, and the demand for food put a premium on the land. The Tudor enclosures had only affected a limited area, but now there was a far greater demand for new and undeveloped land, and also for the deteriorated land on which poor crops and poorer cattle revealed the need for capital and manure.

In the decades of the eighteenth century before the advent of mechanisation, enclosure Acts were few; in Anne's reign there were two, under George I sixteen, and under George II there were 220, but in the latter part of the century, when George III reigned, there were 3,554. In the 50 years before George III, 337,876 acres were enclosed; at the end of his reign 5,686,000 acres had been enclosed.

As in Tudor times, there was great improvement of the soil in the enclosed areas. Robert Bakewell (1725–95) transformed raw-boned cattle and lean sheep into animals twice the size. From 1776 on, Thomas Coke of Norfolk proved the capacity of mixed farming to carry three times the livestock and to produce rich crops of wheat in place of scanty rye. Turnips were grown as winter feed for cattle, and clover to improve the soil.

Earnest farmers followed these examples. Nevertheless, the main impulse that led to the enforcement of enclosures was the opportunity given to acquisitive men to rise quickly to great wealth. It was this that gave the movement its brutality and the character of a civil war between one section of the people and another. Though the swords of the fortune hunters were sheathed in legality, they were none the less keen when unsheathed and so, says Hasbach, enclosures were 'changed into a national curse'.

It was the wealthier inhabitants of rural areas who appreciated the opportunities of seizure, and it was therefore squires, parsons and lawyers who benefited, and became a new and well-established class of land-owners.

Though farms owned and operated by peasants survived in a few parts of England, in general, Green says,

> 'yeomen farmers and peasant proprietors ceased to exist; they drifted to the towns, and sank to the status of workers earning a daily wage. Not only small holdings but also the lesser tenancies gradually vanished in a universal system of large estates and farms.'

The agricultural labourers in this period reached the nadir of their fate. They had no protection from the Church and the monasteries, as they had when Catholicism was the religion of England; their cottage industries had been supplanted by the new machines of the towns; and the days of a trade union for agricultural labourers were yet to come.

They were utterly helpless and hopeless. They were not even slaves, ensured of receiving board and bed from their masters. The landowners ceased to pay wages in kind – in other words in food – because food fetched higher prices in the towns, and the yeomen who had once filled the village markets were no more. Their food was almost confined to wheaten bread, which, being wholemeal, supported life. Their wages were miserably small; so small that the parishes often had to add to the pittance an allowance apportioned from the rates. Because of this, the parish authorities hired out the labourers.

Sometimes, wrote Ernle,

> 'the paupers were paraded by the overseers on a Monday morning, and the week's labour of each individual was offered at auction to the highest bidder.'

The labourers presented heart-rending pictures to their bravest champion, William Cobbett. Here is one, taken from his *Rural Rides* (1821).

> 'The labourers are miserably poor. Their dwellings are little better than pig sties, and their looks indicate that their food is not nearly equal to that of a pig ... The land all along here is good. Fine fields and pastures are all around; and yet the cultivators of these fields are so miserable ... When I see their poor faces, nothing but skin and bone, while they toil to get the wheat and the meat ready to be carried away to be devoured by the tax-eaters, I am ashamed to look at these poor souls and to reflect that they are my countrymen, and particularly to reflect that we are descended from those amongst whom beef, mutton, pork and veal were the food of even the poorer sort of people.'

This degradation of labourers on the land was essentially an English phenomenon. It did not happen, for example, in the Netherlands. Nathaniel Kent travelled in the Netherlands, and, in his *Hints to Gentlemen of Landed Property*, 1775, he tried to awaken his readers to this fact.

In the Netherlands, he wrote, there was an astonishing quantity of provisions, and he recorded that the holdings were all small and the cultivators enjoyed equality.

This degradation, therefore, only happened in England. And even then, strangely enough, it was not inevitable everywhere in England itself. That indefatigable traveller on behalf of agriculture, Arthur Young (at one time a zealous champion of enclosures, but later of the opposite opinion), discovered 'with great delight the life of the small proprietors of Axholme'. (*Report on the Agriculture of Lincoln*, 1799)

An important fact about these small proprietors of the Isle of Axholme was that they were not English, but Dutch. They were a piece of the Netherlands that had been transplanted to England. Their ancestors had arrived on the Isle of Axholme more than a century before Young visited them. The Isle was a swampy property of 46,000 acres between three rivers in Lincolnshire, and had the good fortune of belonging to one of the most cultured and educated men of his time, Charles I. Charles had knowledge of the small holders of the Netherlands, and he brought some of their families over to England to drain the Isle of Axholme and cultivate it. They were true intensive peasants and family farmers who, Hasbach wrote, took every small advantage, cultivated every corner, had the help of their wives, and brought up their sons in their footsteps. He wrote:

> 'They serve the land in the way it should be served; never stinting themselves and as absorbed in their service as any priest in his religion.'

So these peasant families caused Axholme to flourish, and it was still flourishing when was seen by Arthur Young, at the time of the degradation of the small English proprietors and their expulsion by the enclosures.

Axholme is still flourishing today. Sir Rider Haggard in

his *Rural England* (1906), described

> '...its almost inexhaustible richness... it will produce magnificent crops of wheat, potatoes, celery, or whatever it may be desired to grow.'

Years later, Gilbert Slater, in the *Making of Modern England* (1934), seeing heavier crops than he ever saw elsewhere, drew the conclusion that the farmers of the Isle of Axholme had been totally vindicated in their stout refusal to submit to enclosure in the eighteenth century.

> 'Not only are the open fields of the Isle of Axholme exceptionally well cultivated at the present time, but the island serves as a training ground in practical and effective farming, and men who begin as labourers there frequently become large farmers elsewhere.'

These skilled and independent men met with strong resistance from the English farmers who tried to expel them, but they had inherited a tradition of protecting and feeding the soil, which gave them great faith in their own work. They knew the superiority of their methods, and they have not changed. The 14th edition of the *Encyclopaedia Britannica* states:

> 'Their ancestry affects the physical appearance and accent of the inhabitants of the present day.'

On the other hand, the English labourers of the early part of the nineteenth century had lost all courage. They were an unprotected proletariat. In the times of their prosperity and independence, says Hasbach,

> 'they had avoided early marriages and abstained from multiplying as a mere proletariat does; whereas now all such evils appeared.'

This, he goes on to say, with great significance to all narrow-visioned reformers who wish to increase a population, is the response to Malthus, who failed to recognize the psychological elements – despair of the future and of freedom – that lead to a rapid increase of population. As Hasbach put it,

> '...the error was immense'.

Hasbach places the beginning of the slight recovery of the English agricultural labourer at 1834, when a Poor Law stopped the parish allowances which advantaged the farmers, and instead made the farmers pay the whole of the labourers' wages. Actual paupers were put in the workhouse.

But the real betterment, he found, was due to two things; allotments and trade unions.

About this time, many generous farmers gave allotments of land to their labourers for their own use, and were pleased to find that, instead of making them work poorly on the farmers' lands, they worked better. The eternal truth – that everyone likes to take pride in his own work – was demonstrated yet again and, through being proud of the crops they raised on their own land, these humble families now took greater pride in the crops they raised on their masters' land. They did so well on their private land that when a government report in 1843 pressed for the extension of allotments by law, the farmers complained that they had difficulty in getting enough cheap manure, as the labourers wanted it for themselves. The labourers in a very small way were, in fact, turning the scales against the old lords of the manor who had started all their troubles by stealing their land's nutrients through enclosure.

In 1872 the labourers, under Joseph Arch, started a trade union, and

> '...considering the character of the labourers and their natural isolation, they were at first very successful.'

But their efforts to get better wages were defeated by the farmers, who summoned unemployed workers from the towns and impoverished Irishmen for harvesting, hop-picking and other unskilled work in the busy seasons.

After a long period of depression, the unions sprang into life again in the year 1890.

The unions now went to the root of the matter in their attempts to free land from the dominance of money. They supported the Land Restoration League, which wished to put a tax upon rent and increase it progressively until it absorbed and eventually abolished rent, and thus achieve the aim of Henry George. Agricultural and urban unions began to work together to prevent town labourers from frustrating rural strikes, and vice versa. Though poverty, ignorance and isolation of their members kept the rural unions back, they always gave expression to the labourers' desire for land.

Allotments remained the most recognized form of relief. In 1889 a Parliamentary Committee on small holdings, with Joe Chamberlain as chairman, reported, with farsightedness and objectivity, that a well-to-do peasantry was beneficial to any country, nationally, socially, and economically.

This view was supported by the Central Chamber of Agriculture, which maintained that, whereas large farming was suitable to sheep and corn, small holdings were suitable for other types of farming.

> 'The theory that the agricultural population in general was inevitably attracted by the towns cannot be seriously maintained... The labourers did not depart from areas where allotments could be obtained, where good houses could be had at a fair price, and where some independence thereby was theirs.'

The younger generation began to show themselves discontented with village life. The old semi-feudal relationships of the English village were no longer pleasing to the younger generation, who were more willing to migrate to the larger towns, because on the land there was so little chance to raise themselves socially.

Hasbach ends with a review of the labourer from 1894 to 1906, and in these last pages the light of hope is dulled. The prospects of betterment were not there. The generation that was content with allotments, good wages and decent cottages almost died out.

> 'The new generation altogether despises the position of an agricultural labourer... He is at the bottom of the social scale, and knows it; whereas in a town a man can lose his identity among the masses of the inhabitants.'

As a result of his study, Hasbach came to the belief that little or no permanent improvement in the lot of the labourer had been attained. He could not avoid the impression that

> '...in spite of the talk of better wages, the lot of the agricultural labourer in many parts of the midlands, south, south-east and south-west, where often the houses are wretched and both allotments and small holdings are wanting, is such that he is strongly induced to turn his back on the land, even though his sense of self-respect is comparatively undeveloped.'

Even though the labourer strives for a humble independence, it is definitely the end of many people when the proletarian class is placed at the disposal of the farmer.

No statesman had arisen capable of viewing the picture as a whole, or of estimating the total probable result of any measure.

> 'Hitherto failure has attended all attempts to apply to the problems of agricultural labour the principles which have been effective in the realm of industrial labour.'

The consequence has been the demoralization and depopulation of the countryside. Facts confirm that the system of the large farms cannot meet the crisis. Hasbach's final advice is the greatest possible extension of small and medium-sized holdings.

So ends this most instructive and unique book.

Between 1906 (when Hasbach ended his story) and the present day, England has fought in two world wars. In both, her people have been aroused as to the perilous state of their food supply; and in both they might (and almost certainly would) have been starved into submission, had it not been for supplies sent to them by the USA.

In the first war there was a wise increase of allotments so as to increase the supply of food. Powers were given to local authorities to acquire land by compulsion for allotments, and their number leaped from 130,536 acres to 1,330,000 acres. In the interval of peace that followed, however, much land went out of cultivation.

The great efforts made to increase the production of food before and during the Second World War are too well known to be recounted here.

How far we have become separated from the knowledge of how to feed our soil, and how it can best be cultivated, these two great crises have revealed.

In no country is reconstruction by way of the soil more needed than in the British Isles. We have a large population; and we need improved fertility of the soil to render our population safe and healthy.

We need to free ourselves from robbery of the soil.

11

Primitive Farmers

THE WORD *PRIMITIVE* is defined by *Annandale's Concise Dictionary* as 'characterized by the simplicity of the old times'. This definition is an excellent description of the 'primitive' peoples of this chapter and of the two that follow it. 'The simplicity of old times' fits well, for the dictionary further informs us in the entry for the word *simple* that it derives 'from a root meaning one or unity'.

We can now paraphrase our chapter title of *Primitive Farmers* as *Farmers Characterized by Unity*. We must do this quickly, before going on to read other definitions of *simple*, for we shall find that one of them is 'easily intelligible', and yet farmers characterized by unity are not a bit easily understood by modern peoples. In fact, it is because they have so rarely been understood that so many troubles have befallen them as a result of their contact with modern civilizations.

The primitive people to be considered in this chapter are the Kikuyu of East Africa, for a very rare kind of book has been written about them. Its author, to use a phrase coined by Robert Louis Stevenson, eavesdrops at the door of the hearts of the people she describes. Elspeth Huxley, in her book *The Red Strangers*, tells the story of the minds and hearts of the Kikuyu, to whom the British were the 'red strangers'.

Before the coming of the British, the Kikuyu cultivated the land under a system of family ownership. The land was cleared of forest, and farmed. When its fertility was exhausted, a new clearing was made, and the old one given a long rest and allowed to return to jungle. The Kikuyu grew fruits, beans, peas, millet, sweet potatoes and other food crops. They also kept goats and cattle. The fields were worked by the women, while the men protected the fields against the inroads of wild animals, tended and protected their domestic animals, and acted as warriors when young and councillors when old.

They adapted their life cycle to their conditions, which in turn they modified to their own advantage, but to which they did no permanent destructive harm.

An important feature of the tribe with regard to its eventual contact with Western civilization was that it had no metal coinage, nor did it have any other form of durable money. Its currency was domestic animals; the smaller currency being their goats, the larger their cows. In this matter of currency, therefore, they resembled the early ancestors of Western civilization, whose word for money, *pecunia*, was derived from *pecu* (cattle).

In looking at the hearty and cheerful Kikuyu, as they were when contact first occurred, Westerners saw a people who still possessed characteristics reminiscent of the original Latins from whom their own civilization had derived.

Goats, then, were the pecuniary units of the Kikuyu. A poor man had a few goats, a little land and one wife, while a rich man had many goats and fields, together with more than one wife to work the larger possession of land, and more sons to tend the more numerous animals.

Cows also were symbols of wealth; a cow was valued at about a dozen goats. If a man procured the consent of a woman to marriage, he had to pay some such sum as 35 goats, or two cows and ten goats to her father, and sometimes rams and brews of beer were also part of the payment. A field was valued as being worth so many goats or cows. A crime was expiated in a payment of goats to the injured party, or in the case of murder, a fine of over 100 goats could be paid by the clan of the slayer to the clan of the victim.

Goats possessed a second quality of money, over and above their general distribution; they helped a family in times of hardship. Goats are distinguished amongst domestic animals by being most able to feed themselves under adverse circumstances. In a drought, when other animals perish, goats tend to survive. They do this, it is true, at the expense of reduced herbage. They are, amongst animals, those most able to strip vegetative cover and promote erosion and desert-making, for not only do they bite close, but they are nimble climbers; they can quickly strip a hillside and find sustenance in coarse or weedy vegetation.

So they increase and perpetuate the disaster of drought, as does money, when as debt it adds to and perpetuates the seasonal disasters that beset Western farmers.

In 1898, the Kikuyu were first visited by the 'red strangers', as they called the British, and in 1902 at Nyeri, their elders surrendered their freedom.

They were forced to do this by the 'magic' of the British, which was beyond all that they had imagined. A mere noise could kill a man many fields away. The Kikuyu magicians strove to oppose it, but they were as feeble against it as were the prophets of Baal against Elijah. The story itself is, indeed, almost magical, in that an established wisdom, that which had maintained the people so well in a productive cycle of life, should be so quickly dispersed by a mixture of saltpetre, sulphur and charcoal.

Neither the Kikuyu nor Britain can answer the question of why wisdom gets no immediate support from nature.

What is certain, however, is that nature in its own time passes down its verdict, and it writes it upon the soil. In doing so, nature makes itself the final measure of wisdom, and always gives a verdict in its favour.

The Kikuyu cultivated the southern slopes of Mount Kenya at an elevation of 4,000 feet, with a climate in which the British could make their homes. So, with scarcely any preliminaries beyond the display of their 'magic', the British announced that the land which the Kikuyu regarded as theirs now belonged to a distant king.

The Kikuyu, upon the ridges of the hills, had their enemies; the Masai of the plains. The two tribes had fought mainly over cattle. The first thing that the British brought about was peace between the Kikuyu and Masai, but it was not a peace that was the counterpart of war, that is to say, a peace between plumed warriors. The British peace was so odd as to be almost inexplicable.

The men of Kikuyu were commanded by the British to go amongst the Masai peacefully, and to carry the possessions of the Masai, while the Masai themselves – men, women, children and beasts – were ejected from the land of their fathers and sent to a new land. Under the aegis of the peace, the two peoples met and mingled, but in abject humiliation.

As the younger men were deprived of their pride and privilege as warriors, so also their elders found their dignity stripped from them. Whereas it had been their right as councillors to dispense justice and compel the guilty to pay fines to the injured, it was now the British who took over the dispensation of justice and the imposition of fines. These fines now had to be paid not in goats but in money, and the proceeds went not to the injured parties but to the British, who kept the money for themselves.

This clearly was not justice, but theft. There was no

attempt to maintain a balance by means of compensation. The British alone benefited, not only by keeping the proceeds of their 'justice' system, but also by forcing the guilty to do paid work, so that they might earn the money with which to pay their fines.

Later came new and terrible demands. The men of the Kikuyu were taken from their homes and brought down to the sea, which they saw for the first time. They were put into a wagon that rested on the sea and locked into a room with iron walls, the floor of which, when the wagon moved, rocked under their feet. They were overwhelmed with fear; it was like being in the belly of an animal. They were brought to a strange land, where again they carried loads as porters and served the 'red strangers', whose king was engaged in a very big war. They endured hardships so severe that those who eventually returned to their home could not speak of them for many years.

On their return, some of them did not go back to their original homes, but went to take up new land, where the British were also installed. At least now they were free, and able to live on farms of their own making.

But soon, and quite unexpectedly, a red stranger arrived and told them he had paid coins to the *serkali,* or government, and because of this, all the land and even their farms were his. But he did not, he said, intend to take away their farms or their animals. These the Kikuyu could continue to cultivate, but the men must also work for him. They would work one month for him and get six coins or rupees for the work, and then one month for themselves, and so on, throughout the year.

Under this arrangement, large fields of maize were grown and many beasts were pastured for the stranger, and the Kikuyu kept their farms in cultivation and received money.

The early results were surprisingly good. They got their silver coins every second month and what was more, the British had access to markets where they could sell the surplus products which the land produced so abundantly.

So money began to accumulate. An odd thing, however, happened. It was the government who gave out the coins, and the government would not let the farmers keep all the coins they got, but asked for some of them back. As the government themselves made the coins, this was another insoluble puzzle. But, though some were given back as tax, there was still a good number left, either to be buried in the floor of the hut, or to be put in the post office to be spent, when opportunity occurred, on taking up more land and a second wife to work upon it, and more goats for pasture.

So, under the leadership of the red stranger, who had now become for many a friend of sorts, riches – that is to say land, wives and goats – became more plentiful, and the future held out promise as never before.

Then something happened that neither the old ways nor the new magic, even with its new coins, could avert. There were two years of drought, terminating with plagues of locusts and famine. The government sent food to the people, saving many from starvation.

There followed a season which seemed to deliver all the rain that should have fallen in the two previous years. The crops were now not burnt up, but drowned. Furthermore, in spite of the great food shortage brought about by the drought, when any surplus product was now taken to the market, instead of getting good prices, for some inexplicable reason the prices were so low that they did not cover the cost of cultivation.

There followed a further season of drought, with the sun beating down day after day upon the land. The lake in the valley sank to a level unrecorded in living memory. The pastures, stripped by locusts, turned to powdered earth, and dust devils whirled across the valley. Erosion set in. It was as if the new treatment of the earth made the soil take on a new and alien character, causing it to tear itself from its home and flee with the wind. And so the soil itself escaped from the red strangers, something that the Kikuyu could not do.

As has been said, something had gone wrong with the coins of the new currency, and it was now found necessary to debase the currency of the Kikuyu. The British, whom the Kikuyu had had to obey and had come to trust, issued an order limiting their currency in goats, first to ten goats for each married woman, and then five. But this too failed, and the British, finally having no coins left, bid farewell to the Kikuyu and left the territory altogether.

In their place came another, and with him officers of the government. Then came the final blow. All the goats, which in their hunger were eating everything down to the roots, were expelled from the stranger's pastures. The Kikuyu who worked on the large fields were allowed to continue their work, but they were to have no goats; if they wished to keep goats, they and their animals must go elsewhere.

In this way the traditional currency of the Kikuyu peasants, that which had been to them what the coins had been to the British, was as effectively destroyed as had been that of the peasants of India by *Act No. 8* of 1893. It was replaced by a currency which had no relation at all to the local returns of the soil, as the goats had, but was something quite outside the cycles of fortune and misfortune which work and the seasons brought to the Kikuyu.

The new currency, it is true, brought with it certain advantages. In times of famine, it allowed the Kikuyu access to goods far distant from their locality. With it came trade, education, and the creation and improvement of towns as means of livelihood. But it took away something that was an essential part of the life cycle; the wealth of animals on the farms, which rose and fell according to the creative capacity of the soil.

When severe adversity came, animal life was diminished; it was only extreme and rare disaster that had a similar effect on human life. Being a part of the life cycle itself, the value and availability of the currency of the goat moved up and down with the favourable or unfavourable status of the soil. The new coins, on the other hand, had no relation whatever to the soil, local or otherwise; they were completely divorced from it. The new currency had, indeed, nothing at all local about it in terms of either agriculture or origin, and had an existence entirely separate from the life cycle. It was related not to the soil, but to world finance, and as such was the first modern attempt by a group of men to be masters of the world.

Without their goats, the Kikuyu were like the British had been without their coins, and they, too, in their despair, followed the British example. They packed up and left, to return to the ancestral lands of their forefathers.

In her narration, Mrs Huxley describes how the transition of Kikuyu agriculture from a subsistence to a modern capitalist model of farming took as many decades as the same process required in centuries in England, so swift was the tempo of change. Yet despite the different scale of time involved, all the main features of the English example reappeared in the Kikuyu story.

The families returned to their homeland, confident that according to tribal custom they would have a right to the land which their fathers had cleared. On their arrival, however, they found changes as great as those they had experienced in the land they had just left.

Relatives had taken over both the land and the glades.

The glades had been turned into pasture, and had gained something that had been unknown in the past; fences enclosing them. Previously, all pasture had been open, and was common ground held by all the villagers. Fields in the past had only ever had temporary fences to protect crops from wild pigs and other animals, but the fencing they now saw was substantially made and permanent.

The cultivation of the fenced-in fields was also different to that which they had expected. The native method of hoeing by hand had been supplanted by a plough with oxen to draw it, and they soon discovered that there were other new methods of cultivation as well, including the rotation of crops. Still more surprising was a square house built of stone, with windows, a veranda and an iron roof. Surrounding the house was a garden with flowers, and fruit trees planted in rows.

The family would have looked about for goats, but there were none to be seen. The goat, once the unit of currency and also the victim of religious sacrifice – and so in two aspects closely interwoven with Kikuyu life – had, under British influence, been discarded.

There were other measures of wealth as well, and the returning Kikuyu soon realized that what they had seen of the British in the land they had left was repeated here. They were looking, not at communal or tribal land any longer, but at the estates of British colonialism. They also found that their own kinsmen and relatives had become like the British in claiming, under the authority of the government, that the land now belonged to them. In short, they were looking on something new; the notion of *private property.*

Certainly their relatives had benefited greatly through the actions and policies of the government, and by listening obediently to its agricultural officers. In its role as an instrument of progress, the government had created something new to the Kikuyu, though not, of course, to the English – they had introduced the notion of the 'lord of the manor'. They had made some of the Kikuyu local land-chiefs, some of whom had become so rich that they now had many wives to serve them.

Even the form of the wives' service was new, for it was they, not the men, who now tended the cattle. There were surely sons enough for the work, but they had all been sent to the government schools, and this placed them above tending cattle. Education took the young men from the land to the towns, where they became clerks, teachers or policemen, or took other positions of service to the government. In these services there lay safety, for in the towns, during periods of drought and famine, these younger people still had enough to eat, still travelled comfortably in buses to work, still dressed in European clothes, and danced in European fashion. The great affliction of the countryside was not visited on the towns.

The returning family saw and heard all this. Particularly, of course, they noted what concerned them most; the stone house, the rows of fruit trees, the cattle, the fencing and other changes on the land that according to custom was theirs.

On the one hand, then, was their traditional right, and on the other was the new and inescapable fact of private ownership. The father, now an old man nearing his end, wished to bow before the power of the new, but the son was unwilling, and prevailed. And so a claim for the land was lodged by the family.

The case aroused the keen interest of the whole locality. It epitomised the conflict that was everywhere occurring between the old and the new. The elders stood firmly for the tribal laws of inheritance and the safe living upon the land which they gave to each family, and so they opposed the new rights which made men dependent upon the will or whim of so-called owners of the land.

The younger generation stood as firmly for the cousin, because of the improvements he had made under the guidance of the government experts. This, they said, made the land his. As for the family, if dispossessed, there were other ways of making a living available to them, such as becoming labourers on the roads or railway, or in house-building, or porterage, or even in Nairobi, by working as taxi or bus drivers. They could even stay on the land as hired labourers, receiving wages from the new owners.

As the claimants could not afford to pay compensation for the improvements, the land was finally awarded to the cousins who had taken possession in the family's absence. But the claim of the family was also acknowledged, and land belonging to the clan, of size and quality equal to the original clearings next to the forest glades, was awarded to them.

So, after many sudden and unpredictable changes of fortune, the family attained once more the traditional security of the homeland. But even here, they had to submit to the effects of what was to become by far the most dangerous change of all.

The old father died and he left behind him a legacy; the prophetic pronouncement of his recently deceased friend Irumu, who had been the seer of the tribe:

> 'When women walk all day to seek firewood and when cultivation lies naked under the sun, then shall evil come. But on the days when trees again darken the ridges and bring shelter to the weary, then shall good fortune return.'

From the deep, inward oneness with the local life cycle which such tribal wisemen possess had arisen a true vision of the coming of the 'great erosion'.

Where the new greed for land as property had caused

too many trees to be felled along the ridges of the hills upon which the Kikuyu had their homes, the effect of the torrential rains was unchecked. It would pound the top soil into mud, which then ran down the hills. This was the beginning of water erosion, which as it spreads forces women to walk further each day in search of firewood.

When the fields were broken open by the plough instead of being lightly stirred by traditional digging knives, and when they were made to grow just one crop rather than several crops of different heights, foliage and roots, the cultivated land soon lay naked under the sun. A dry season could then turn the soil to dust, and some of it would be blown away by strong winds. This was the beginning of wind erosion.

These two erosions form the last part of the story of the entry of the Kikuyu peasantry into modern civilization. There were greater demands on the fertility of the soil, as the native urban population increased under the early rule of the Westerners. Many new ways of earning the new coins appeared. The colonial government demanded more coffee, sugar, cotton, hides, maize, sisal and other commodities for export. More land was turned over to cultivation, its fertility taken up by the crops, while the rule of return was of course neglected.

Here is an account of the last phase of this process as it affected the Kikuyu, as described by Jacks and Whyte in *The Rape of the Earth* (1939). Their account completes the story begun by Mrs Huxley.

Erosion, they write, has attacked the lands of the Kikuyu, due to agriculture being forced at too speedy a pace through the desire to obtain cash from the sale of crops, and also because of the demand for more food crops by the increasing population, most of which migrated to the growing towns.

The original family-based mixed farming for sustenance succumbs to the new methods of commercial farming. One farmer will concentrate on the growing of maize, another will stock – or overstock – the land as pasture; both are conducting a rape of the earth. They farm for cash, and not heeding the rule of return, they consistently take more fertility than the soil can recurrently yield. They treat the soil as if they are its conquerors, not its partners.

As part of the quest for more crops, the Kikuyu cultivate not only the ridges upon which they had their homes, but also the easier slopes of the hills. There is soon a loss in the quality of the soil, a loss of that wonderful air-containing, loose adhesiveness of the soil due to good humus. With this degeneration the natural elements of the rain, wind and sun, once friends and partners, now, at the times of their greatest strength, become enemies.

The government has taken no proper measures to prevent this. There is a 'lack of conservation measures in general', say Jacks and Whyte. The European owners mostly exhaust their estates by the same disregard for the precepts of nature.

> 'In the European areas, erosion is caused by exhaustion of the soil through long continuous cropping, without the adoption of methods to prevent erosion and maintain the humus content of the soil. The results of land misuse are only now becoming apparent in a grave form, as much of the land in the settled areas has only been cultivated for fifteen to 25 years. Some areas of Kenya have already reached such a state of devastation that nothing short of the expenditure of enormous and quite impossible sums of money could restore the land for human use above a bare and precarious subsistence standard...

> Generally speaking, erosion has become serious only during the past five years. In addition to the causes listed above, the invasions of locusts of 1929–31 and the drought of 1931–5 greatly accelerated the process, and were largely responsible for making it so apparent in the space of just a few years.'

The English came to the land of the Kikuyu because although situated upon the Equator, it is highland, and has a climate in which the English can live and farm. They make their homes there, but to maintain their accustomed standard of living and to save money, they concentrate on farming for profit, and to achieve this they follow familiar methods of modern farming.

That the fertility of soil is exhaustible and also that the methods used (which in the cool, wet climate of Britain led to the soil being slowly but surely depleted) will, in Kenya, deplete it at an even faster pace. That the sun, wind, rain, goats and cattle – which were all natural parts of the old life cycle – should become enemies rather than friends is foreign to local traditional experience and knowledge. The government, wishing the Kikuyu to share in the wealth that the new methods can create, induces them to adopt the new values.

The intentions were good. Both the white and black populations seek to profit through progress and science. Though the English claim the land as belonging to their distant king, any further exploitation of the Kikuyu is not the king's wish. In July 1923, His Majesty's Government decreed that the interests of the natives must be paramount over those of all immigrants, including the British, and that on no account were the interests of the black population to be scarificed for the advantage of the whites.

The English, who had announced to the Kikuyu that the land had become theirs through the payment of money, nevertheless soon became their friends.

But both the black and white races depended on the soil, and it was the soil that was soon sacrificed. It was stripped of its cover with eager haste and a tragic lack of understanding. The final result is not yet known, but what is known so far is sufficient.

In some parts, the prophecies of Irumu have been fulfilled: the women walk all day to seek firewood and the cultivated areas lie naked, exposed to the heat of the sun. Can then the future days of which he spoke also cease to be mere words, and be realised in fact?

> '...on the days when trees again darken the ridges and bring shelter to the weary, then shall good fortune return.'

Money, which has been the root of this evil, is unable to rectify the situation. Only enormous and quite impossible sums of money could restore the land. And before money can enter the picture, of course, there must occur a fundamental change of values, a change of outlook and a change of faith.

Nature is very careful, but men are careless. For example, in some of the species of acacia trees in Australia, the leaves are suppressed, and the leaf stalks or petioles are vertically flattened and assume the function of leaves. The vertical position of these petioles prevents injury from excessive sunlight, as with their edges aligned to the sky and earth, they are not so exposed to the light as are horizontal leaves. Scientists may be able to explain how this comes about, but to the layman thinker, it is an exquisite example of nature's care and design. Such an example should impress farmers, telling them:

> 'Do likewise. Exquisite care is necessary in the preservation and adjustment of the details of life cycles. *That is what farming should be.*'

12

Nyasa

HERE IS THE STORY of another traditional farming people of East Africa. They live about 1,000 miles south of the Kikuyu, and occupy highlands of lesser elevation, at about the same distance from the sea.

In 1935, the government of Nyasaland became concerned by the increasing exodus of able-bodied peasants from their homeland, and the Governor appointed a committee of inquiry. How alarming the exodus and its far-reaching consequences were was soon revealed to the committee. As they travelled and saw and questioned, the extent of the tragedy was disclosed. Their report included the following:

> 'We must confess that, six months ago, there was not one of us who realized the seriousness of the situation. As our investigations proceeded, we became more and more aware that this uncontrolled and growing emigration has brought misery and poverty to thousands of families, and that the waste of life, happiness, health and wealth is colossal.'

This statement is worthy of close attention. At the outset, it should be noted that there was not a single rural native on the committee; it was assumed that wisdom lay outside the land. Accordingly, the committee men were not rural men, and had very little knowledge of the rural life of Nyasaland.

Consequently, within six months, they found themselves astonished – even overwhelmed – by the disruption of the rural life cycle for which they and their kind were responsible.

The wealth of which they wrote in the above quote was the wealth of the land. They further explained their position:

> 'We consider it essential that the whole Protectorate should be surveyed by local agriculturists with the idea of determining the best uses to which the land can be put, regarding the land not as something to be exploited piecemeal, but as the sole capital of the Protectorate.'

The language, one will note, is that of money-minded men; the land is called the *capital* of the Protectorate; as capital it must not be exploited for industrial profit, but put to the best use possible as the only means of livelihood for the people of the Protectorate.

How this was to be done was to be decided by local agri-culturists. This term did not include native farmers, because they would not be capable of conducting the surveys that could provide the necessary facts and figures from which to draw conclusions.

The committee was not seeking a renovation of the traditional indigenous social and farming life, from which improvements could eventually develop, but instead considered the question from the modern Western agricultural viewpoint. In this way they sought to decide how to put this particular 'capital' to its best uses.

The indigenous methods by which the people of Nyasaland farmed resembled those of the Kikuyu. They cleared a part of the forest and cultivated it for as long as it gave good results. Then they abandoned it for a number of years during which, through the encroachment of the neighbouring forests, it reverted to its matural condition. This is called 'shifting cultivation', about which Jacks and

Whyte say:

> 'Shifting cultivation, although it kept men as unimportant servants of wild nature, maintained soil fertility indefinitely, since the forest drove the cultivator out and re-assumed its beneficent control as soon as any sign of soil exhaustion occurred.'

The indigenous method, therefore, included as a practice, if not as an intellectual precept, the indefinite maintenance of soil fertility. The Western agriculturalist land owners only awoke to the devastating effects of the loss of soil fertility after it had occurred to a significant degree, and only then devised methods of preservation.

All this occurs because money is antithetic to real traditional and conservative farming. The whole conception of money plus interest is foreign to the soil. As we have seen previously, when money is lent, it is in the expectation of getting not just itself, but more than itself in return, the additional sum being *interest*.

A crop, on the other hand, does not produce anything more than what it gets from the soil and the air. *The creative power never creates anything extra*; it merely changes forms. In nature there is only transformation, not addition.

The conception, then, that money can produce extra money –something over and above itself – is not one derived from the creative power of the soil or indeed anything natural. That perhaps is the ultimate reason why the practice of applying interest to loans has been so strongly condemned by religions and philosophies. Peasants feel it to be wrong, and the poets, who according to Dante are 'those who know the secrets of nature', also know it to be against nature and unreal, and therefore both inimical to the intellect and morally wrong.

For these reasons, money-centered farming, however

scientific, cannot create the honest constancy of equivalent return. It strives to get more than it gives, and therefore eventually creates a situation in which survival itself becomes uncertain. This, then, is the story of Nyasa; a government dedicated to the money system demanded that a traditional farming people undertake money-centered farming.

The traditional people belonged to a complete and natural life cycle, in which surplus crops were exchanged for other human needs. There was no space in their work or habits for anything over and above this life cycle; nothing, that is to say, which could be stored away as dead capital, or discarded as not wanted. It would be turned immediately into wealth, which meant cattle or other such visible 'goods'. It certainly could not be symbolized and banked.

Consequently, when the Nyasa government demanded a hut tax to be paid in money, it drew the peasants into the money system without any preparation or aptitude for it, and without defence.

They could easily have paid the tax in goods or kind, according to their custom. But they did not have the coins which the government demanded, and their own elder men, because of their need for money, ordered the younger men to pay for their brides in coin, rather than cattle.

The cash asked for by the government through taxation, wrote the committee in their report, was considerable. It was more than a farming district, after providing for subsistence, could earn. The committee described, as examples, five districts which had to pay taxes of £18,000, despite earning only £14,000 (market earnings of £1,000 and wages of £13,000). In response to these urban demands on primitive farmers, therefore, none of them proved to be fit to survive on the land of their fathers.

There was only one thing to be done, as was done

in England by so many of the harassed Tudor and Georgian peasantry; they had to evict themselves and seek employment in the modern mining towns of Tanganyika, Rhodesia and the Transvaal. Hence, out of a total population of 1,600,000, there were 120,000 farmers continuously out of the country – 50 to 60 per cent of the able-bodied workers. Basutoland and Swaziland, farther south, had almost the same percentage of young workers absent from the land.

The workers, who had been partners of the soil, were now dispersed like eroded soil to regions where they were made to take from the earth not crops, but the gold that was the god of the money system. 50 to 60 per cent left their farms; yet the government of the Belgian Congo had been advised by one of their committees that even the absence of just 5 per cent of young able-bodied men from an African village would upset the whole economic and social balance of the community.

There is scarcely any need to describe the state of the peasant families who remained – the women, old men and children who strove to carry on the cultivation of the land; the fields overgrown with weeds and jungle invasion; the huts falling to pieces; abandoned fields and crumbling villages. It was as if the Tudor period had cast its dark shadow upon Nyasa and the neighbouring lands.

A partial and unlikely remedy came through the lack of chastity of the Nyasa women. Wearied by poverty and the increasing illness which accompanied it, many gave up any attempt to remain chaste while waiting for their husbands' return from the mining towns, bearing the cash and venereal disease that they had acquired there.

So, when natives of Portuguese East Africa discovered that there were women and land across the border, they seized their opportunity. At the present time, it is said, there are as many, or even more, such male cuckoos resident in Nyasaland as there are Nyasa men.

The story of Nyasaland is one of a colossal waste of life, happiness, health and wealth. It is the tale of the misery of a shattered life cycle. It is not a tale just for the heart, but also for the mind. History has repeated itself again. Post-Punic and Tudor history, with their evictions, poverty, reallocation of wealth and disintegration of social networks, are repeated in the story of Nyasaland and its neighbours. The distances involved in this case are greater, and the centres of wealth that are nourished are in London and other cities, as well as the middlemen who profit along the way.

The story is one of debasement caused by money. The peasants were forced to leave their farms to earn cash to pay the taxes that had been imposed on them. It is a tale of the clash between the demands of the money system and the age-old ways of traditional farmers. The foes were face to face.

The reader may wonder why traditional peoples seem to fare so badly when they come into contact with modern civilization.

Here is one part of the answer: the money system often exercises its power through the use of explosives, bombs, tanks and the other instruments of war. Against the soil, however, its weapons are different, but nevertheless, through them the money system is far more widely and more permanently lethal than it is through the destructive efficiency of its war machines.

It kills at the source – by destroying the partnership between the soil and the peasants. The money system produces eventual desolation; peasants and soil vanish, and with their loss, what was once a source of health and creative power is given over to waste and death.

13

Tanganyika

BETWEEN KENYA AND NYASALAND lies the great, sparsely inhabited territory of Tanganyika. Here there is a life cycle of a very remarkable character, and which contains within it an insect – the tsetse fly.

This insect has come to play the part which the lions of Judah once played as defenders of the natural forest against the intrusions of man. Palestine no longer has its lions and the consequence is that, when one flies over it in an aeroplane, one looks down to see barren rock where there should be forest.

The tsetse's method of defending its forested home is more subtle than the terror by which the lion once kept men at bay. There is nothing regal about the tsetse, but nevertheless the part it plays in its environment is one of the most remarkable to be found in nature.

It feeds, like the mosquito, upon the blood of both animal and human. Tsetse is also a host of the microscopic organism *trypanosome*, which it may inject into the blood of the bitten animal. When it infects animals normally found in its own environment, the animals live; the trypanosomes do not harm them any more than a number of microbes which ordinarily live in humans do harm to us.

But if man's domestic animals and man himself invade the tsetse's territory, it is a very different story. On the expedition to Tanganyika, to which the Kikuyu peasants were taken as porters and endured such miseries, none of the animals imported into Tanganyika in the service of the

British forces survived. Practically all that were not killed accidentally succumbed to the tsetse fly.

In the same way, it can be just as lethal to humanity. The first trypanosomes were brought by cattle driven across the watershed between West and East Africa. The tsetse fly already existed in parts of Uganda, and soon became host to the trypanosome. 200,000 out of 300,000 people died in six years.

Humanity, therefore, has a very great and justified fear of this insect, a fear similar to that of the Palestinians of the past for the lions of their forests. The tsetse evicted them and their cattle from its forest areas. It is said that in the 365,000 square miles of Tanganyika, two thirds of the five million inhabitants have to confine themselves to just one tenth of the total territory.

Then came white men, determined not to be kept at bay by the tsetse. They cut down the trees and bushes near streams, lakes and pools, in the shade of which the tsetse lives. The result was an erosion, not so threatening and extensive as in Kenya, because the area that is cultivated is so limited, but still serious enough to cause an eventual halt in the clearing. It became clear that trees must be left in place to protect the soil against the heavy rains of tropical East Africa; otherwise the forest first became savanna, then coarse grassland and, eventually, if this poor pasture is overstocked with cattle, barren waste.

The hydrological or water cycle, in which vegetative cover plays a vital part, had to be preserved, and consequently the wholesale destruction of the haunts of the tsetse along the rivers and around pools and lakes was abandoned. In its place cautious removal of bushes and trees favoured by the fly is being tried.

Indeed, in no part of Africa has the value of distribution and conservation of the water supply been more thoroughly grasped than in present day Tanganyika. In the Kilimanjaro Native Co-operative Union, which claims 24,000 members out of the 36,000 farmers on the slopes of Kilimanjaro, there are 26 societies. The reason for this number is that it corresponds to 26 streams which have their origin in the great mountain and water its slopes.

Under the guidance of Sir Donald Cameron, geologists, plant ecologists and water surveyors have joined together to educate farmers in the local character of the water supply as a whole. They have marked out the catchment areas of the 26 streams. Each catchment area with the river to which it gives rise has been made into a separate entity, and is presided over by a native chief, and all 26 entities are united in the Co-operative Union.

So, there are 26 catchment areas, 26 rivers, 26 cultivated areas, 26 chiefs, 26 communes – and one Union. It is a real association of communes and the assembly of the Union is a real House of Commons, people sharing a common source of life and not the mockery which similar assemblies have elsewhere become.

So in the strange way in which nature replies to human acts, we have been shown that the tsetse, which has been such a prolific killer of both humans and animals, has nevertheless been a great saviour of the source of life, the soil. Had it not been for the tsetse, the rich soil fed by the greatest mountains of Africa would have been greedily seized and its stored fertility turned into cash, until finally only irrevocable erosion would have stayed further ravages. But, thanks to the tsetse, the soil would not suffer this onslaught; the tsetse has prevented it. In the words of R. Whyte:

> 'the presence of the tsetse in many parts may be a blessing in disguise. It can be regarded as the trustee of the land for future generations.'

The tsetse is a pest to man, but man, greedy and eager to make his fortune from stored soil fertility, is a pest to life itself.

So the strange story of Tanganyika ends with the little tsetse still defending their waterways against the lords of the earth, and so giving time for nature in her own way to tell these lords that, masterful though they may be, if they claim to be the *masters* of nature, they are doomed. Humanity must relearn with humility that we are creatures and students of nature, and, this time, a little insect shall teach us.

14

Humanity and the Earth

BEFORE CONTINUING WITH THE STORY of the present misfortunes of the soil, we shall spend a chapter considering just how connected with the earth humanity is.

We are accustomed to spending a good deal of time thinking through our problems, but we rarely meditate deeply on them, nor do we often bother to distance ourselves from the familiar. We accept the air as air, the sun as sun, and the earth as earth without at any time considering them objectively, or from a distance, so that we can comprehend both them and ourselves in relation to them.

'Earth we are and to earth we return' is a wise and familiar saying upon which we may well reflect. It seems that this earth now under our feet is in some way *us*. To it and its subterranean regions, we and so much else in the world above belong.

The interchange between the visible and the invisible worlds is continuous. The great majority of humanity cares little about what goes on below ground, but since humans are, it seems, unique among the world's creatures in possessing the capacity for meditative thought, we have nevertheless gathered a good deal of knowledge about this planet.

We can scarcely reach deeper than the crust of the earth, but in it we search, in accord with an instinct which tells us that, though we have spirituality, we are still essentially of the earth, and when we search into the earth, we are at the same time achieving a deeper understanding of our own being.

Although we live in the visible world, we are nevertheless destined to return to the earth. Electricity can be separated from the earth and made to run trains, drive ships, illuminate cities at night, and draw great clouds together over thirsty lands, and like man it must return to the earth.

Similarly, modern farmers separate land from its natural state of forest and prairie. They grow products for humanity's use, but in the end those products too are destined to return to the earth.

So also it is with water. Water rises as vapour from the ocean and ascends to the skies, where it takes visible form as clouds. Then it descends again to the earth as rain and takes visible form as brooks, rivers, lakes, ponds and dew.

We humans, too, separate some of it for our own purposes. With irrigation fields are watered, with conduits our cities are watered, and with tanks and reservoirs we provide water for ourselves. Once used, these waters return to the invisible realm; they sink into the earth or into the depths of the ocean, from which once again they will eventually return to the visible world.

We human beings who bring about these transitions are conceived by the sparks that set our being in motion and that spring from the mystery of creation. From the very moment at which the two sparks – male and female – unite, we are terrene, of the earth.

Heredity, in all its variety, comes from two cells so small that it takes a microscope to see them. In these two cells, for us and other beings of the earth, there is something of the magic of predestination. It is they that determine the creation of human, or animal, or plant. In humanity, they determine sex, colour, and character. Though only two tiny specks, they have within them a multiplicity and complexity of destiny that is beyond our understanding.

We may know that there are so many genes in each cell, but to know such details is not to understand the depth of the mysteries they embody.

In this early stage, as in later ones, we receive the means of growth from the earth and from two things which also have an earthly phase, namely the air and water. These means of growth are made up of substances, many of which have been identified as unique entities, and have been described as the *elements*. There are only 90 currently known natural elements, but they occur in so many combinations that we should be entirely lost if we had to manage them ourselves. It is nature that manages them and their interchange.

This we know – if it were not for this combining of elements, there would be no life. Nevertheless, scientists boldly isolate both individual elements and combinations of them, identify them with tests, weigh them and give names to them, and try to force a halt, for such time as suits their purposes, to their constant transitions.

Here are some of the elements that have been found to be a part of the human body: nitrogen, oxygen, carbon, hydrogen, sodium, potassium, sulphur, iodine, fluorine, manganese, silicon, cobalt, copper, iron, zinc, lead, arsenic, lithium, magnesium, aluminium, boron, chromium, strontium, cadmium, barium, tin, vanadium, and titanium. Some of these 28 elements may not be essential to human life, but they are part of it, for all have been found in sewage sludge. They may, one can suggest, be essential, if not to life, then at least to certain qualities of life.

The four great elements of our body, brain, thoughts and affections – nitrogen, oxygen, carbon and hydrogen – are all aerial, as if they have to rise to the heavens for purification before turning back with pristine vitality to the earth. Perhaps there, bathed in the rays of the celestial bodies, they gather that marvellous power of combination which makes them the supreme elements of life. In their endowment of life they show a singular affinity for each other, an affinity so dazzling that it blinds our very thought in conceiving it. They associate together in innumerable patterns, as if in the great spaces from which they come, they resemble Wordsworth's birds:

Hundreds of curves and circlets, to and fro,
Upward and downward, progress intricate
Yet unperplexed, as if one spirit swayed
Their indefatigable flight.

It is these four elements that, joining together in almost infinite varieties, form the proteins of living matter. Some of their steps in the protein dance have been identified by organic chemists. These steps are called *amino acids.*

Here is one (*luecine*), and this is how it is written: six atoms of carbon, thirteen of hydrogen, one of nitrogen and two of oxygen, or $(CH_3)_2$:CH. CH_2. $CH(NH_2)$. COOH. Alternatively, they may be shown as in the diagram to the right.

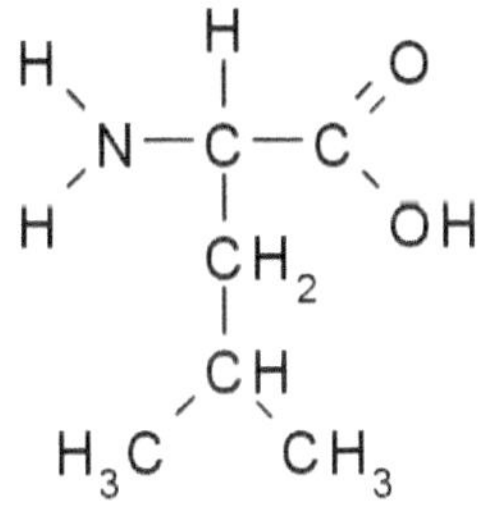

The number of possible proteins is too great to be imagined. Berg gives them as 6,708,373,705,728,100. The transition of associations of elements from one temporary form to another gives one a glimpse of the constant and amazing variety of living matter, before which we can only, with such glimpses as we have gained, regard our own creative and manufacturing power as something – excellent though it may be for us – base and humble before the whirling, form-making artistry of the elements.

When nitrogen steps aside from this quadruple

partnership and leaves just carbon, oxygen and hydrogen, the three again combine in the less dazzling combinations of the carbohydrates, or starchy and sugary substances of living matter. They too are illustrated by the chemists in formations more regular, but nevertheless as wondrous as when nitrogen is involved. Here is a common sugar, *dextrose*:

$$CH_2OH.\ CHOH.\ CHOH.\ CHOH.\ CHOH.\ CHO.$$

If this form were to be placed amidst a number of surrounding mirrors, there would be an equal number of reflections. There are actually sixteen such reflections of the *dextrose* molecule, four of which are found in nature, while twelve have been prepared synthetically by Emil Fischer and others, but have not yet been found in nature.

There is something sober and shapely about the carbohydrates, for the majority of them are so many atoms of carbon in combination with so much hydrogen and oxygen, combined as they are in water, or H_2O.

This cannot be said of these three elements when nature, with her marvellous dexterity, uses them to construct the fats. Here, for example, is an arrangement which makes a fat:

$$C_3H_5\ (O.CO.C_{15}H_{31})\ (O.CO.C_{17}H_{33})\ (O.CO.C_{17}.H_{35})$$

Even with this juggling of the three elements in the making of the carbohydrates and fats, nature is not content. From them, she fashions certain hormones which have governing roles within the body, such as the hormones of the testes and ovaries, and also one found in the adrenal glands which in excess can give a beard to a woman, along with other qualities of masculinity.

Some vitamins are also made from the same building blocks. By adding nitrogen, there result one or two other hormones and vitamins, and yet again with the addition of iodine and nitrogen the hormone of that very dominant gland, the thyroid; and with nitrogen, sulphur and chlorine, we get vitamin B.

When one considers these four marvellous elements, is it strange that humanity, which derives so much of its vitality and the fabric of its spirituality from them, should not from the beginning have felt an intimate unity with the heavens above us? Truly, it seems that we humans have a heavenly as well as an earthly body.

Yet, in its worship of money, in the cities humans shut themselves off no less from the clean air than they do from the earth. Consequently we are less whole and healthy, and must go to the seaside or the countryside to recover some of our health. The authorities even have to plan camps for children and adolescents so they can breathe clean air under an open sky.

We do not know how wide, subtle or deep the total extent of our deprivation is, because our real place within the life cycle is unknown to us and, at best, the subject of only fragmentary research.

How then do we dare to proclaim ourselves the masters of nature and the lords of creation – we who have broken our own life cycle, divided ourselves from its earthly and heavenly elements, and who look to men of mediocre health and physique, sitting in their laboratories, for guidance in these immortal truths; truths that are clearly evident in our inner selves, and are written for all to see on the open face of the great sphere upon which we live?

There are other aerial elements (such as argon, crypton, neon, xenon and helium). Of the relationship that these elements have to life, we currently know little or nothing.

How many of the 28 already named in this chapter are essential has not been determined, but it has been discovered that mere traces of some of them are essential.

Thus, in the case of the black rot of sugar beet, it has been found that this disease occurs if there is lacking a necessary trace of boron in the soil. Similarly, a mere trace of manganese protects oats from black speck. A fatal disease of sheep in parts of Australia and New Zealand is made curable if a little cobalt is added to the soil. In Florida, cattle were found to die until a trace of copper was put in the fields in which they pastured. It is probable, then, that all these 28 elements are workers, and have their roles in the cycle of life.

There is, then, a procession of the elements and, though there is no pause in it, it may be said to start in the microbic and fungal stage, in the soil.

In humanity, the procession starts in our bodies, where the breaking down of waste substances by microbes begins in the lower bowel. Microbes in a healthy situation are friendly; their hostility only appears when living matter lacks what we call 'quality'. When quality is lacking, the microbes set about hastening the return of the living matter which lacks quality – their host – to the soil. This is not a good development if you happen to be the host!

By far the greater part of the microbe world is, then, not only friendly, but it is merely ourselves in a different form; our elements are their elements. They make us and we make them.* Therefore, when we concern ourselves about them, we concern ourselves with what we are. This is one of the secrets of healthy food. If we take elements out of the cycle and disperse them in the sea, we are robbing ourselves.

The microbes then take measures, as it were, to save themselves. Unfriendly microbes multiply. One witnesses, in fact, a break in the *mores*, or the morality, of the

* This statement may be more literally true than the author intended. Interested readers should research Professor Antoine Béchamp's work on pleomorphism and cell biology, along with the work of his successors – Rife, Enderlein, Naessens, etc. – *Ed.*

microbic world. The microbes start exploiting the weak for their own benefit, they become aggressive, send the weak back to the ground, and become sufficiently emboldened to attack the strong. But it is the *original weakness that brings about this break in morality and causes one phase of the procession of the elements to become the enemy of another phase.* The microbe theory and the dominance of money are no strangers to each other; they have been at odds for thousands of years.

It could all be so different. These marvellous elements are like the notes of the piano, which under skilled and reverent treatment can produce an infinite number of melodies and harmonies. In the rhythm and the completeness of the forms they make in the natural world, one can indeed see a wider reflection of that music to which the ancient Greeks gave the highest place in human culture. Misplaced, however, they make cacophony – the hideous noise that now roars throughout the inhabited globe, destroying much of what it touches.

Humanity must revere and respect these elements. We must lose none, we must spoil none. We must consider them wherever and however we meet them as a part of a great being and becoming in which we have our share. Whether as non-farmer or farmer, it should be our goal *to understand the cycles of life and keep to them.*

We should know that, as humans, we tend to be so anthropomorphic, so self-centred, that we interpret food only from our own point of view. We think of it as things of the day, the market and the shop – as bread, vegetables, meat, eggs, fruits and milk; or as things of the factory – processed, preserved, tinned, bottled, dried or dehydrated; or as things of the field – as growing grains, vegetables and fruits. We think of them as things in themselves, as indeed we must do in the course of day to day life.

But to see the quality in our food and also to maintain that quality, we must also think of it as the parts of a whole. This we have failed to do, and it is a significant failure in thought and observation. With that failure humanity has become, in the words of the great seer F. King:

> '...the most extravagant accelerator of waste the world has ever endured. Our withering blight has fallen upon every living thing within our reach, ourselves not excepted.'

The human race pursues the path of collective suicide, while it chants the hymn of progress.

Humanity is terrene, and everything that is of the earth is of importance to us, for we are of and for the earth. As the sugarbeet gets black rot without its trace of boron, oats get black speck without their trace of manganese, and sheep get 'pine sickness' without their trace of cobalt, so we also require such fine tuning for the perfection of our physical and mental health. If we deplete our life cycle, we ourselves are depleted.

In the United States, the depletion of the soil has awakened alarm, and scientists now make statements which seem extreme but may well be true. For example, it has been said that 99 per cent of the American people show some lack of minerals.

Dr Sherman of California has said, as has Sir John Orr, that more than half the population suffers from calcium deficiency. Dr Northen of Alabama has added a number of minerals to the soil and found that, though vegetables and milk produced by it had a normal appearance, they nevertheless had a very different mineral content.

Quite new standards are, therefore, needed.

Textbook analyses, once made, can become widely accepted. Often, however, they adopt standards arrived at through the study of soil that has been injured by faulty practices, so of course they themselves arrive at incorrect conclusions. *We need to adhere to the standards of perfectly healthy soil.*

Man's bodily substance, when not lost to the sea, returns to the earth many times in the course of his life. The saying 'Earth thou art and to earth thou shalt return', used in speaking of the dead, is no less true of our living bodies.

We are animals of the earth. We cannot escape, and so we live as a product of the soil, either conserving it or depleting it as we do so. At present, modern civilization depletes it relentessly. The story of this depletion is in its way both mystical and inexplicable, and ultimately it is one of retributive justice.

The old doctrine that sickness and wars were the punishments of God seems to have an element of truth to it. It seems that, in non-recognition of it, man acts with a perversity little short of insanity.

15

Sind and Egypt

ACCORDING TO A FAMOUS SAYING, the greatest benefactor of mankind is one who makes two blades grow where formerly only one grew. This aphorism could well serve as a description of *perennial irrigation.*

This chapter is concerned mainly with a land where one of the world's largest schemes of perennial irrigation is found – the province of Sind, in India. The Lloyd, or Sukkur, barrage which controls this scheme was opened in 1932.

The creation of soils from the weathering of rock is a process which takes a very long time. The geologist T. Chamberlin, in an address given at a 1908 conference of the State Governors of the USA, spoke of the tremendous importance of the conservation of the soil:

> 'We have no accurate measure of the rate of soil production. We know it is very slow. It varies with the kind of rock...
>
> Without any pretensions to a close estimate, I should be unwilling to name a mean rate of soil formation *greater than one foot in 10,000 years* since the glacial period.
>
> I suspect that if we could determine the time taken in the formation of, for example, four feet of soil, it would be found to be at least 40,000 years.
>
> Given such an estimate, to preserve a good working depth, surface wastage should not exceed such a rate as *one inch in a thousand years.* Even if one chooses to accept a more

> liberal estimate of the rate of soil formation, it will still appear, under any intelligent estimate, that surface wastage is a serious menace to the retention of our soils under our present management.
>
> Historical evidence reinforces this danger. In the Orient there are large tracts almost absolutely bare of soil, on which stand ruins that suggest the historic presence of flourishing populations. Other long-tilled land bears similar testimony.
>
> It must be noted that more than just the loss of soil fertility is threatened. It is the loss of the soil itself, a loss almost beyond repair. When our soils are gone, we too must go, unless we shall find some way to feed on raw rock, or its equivalent.'

This is a succinct description of the end result of unwise cultivation of stationary weathered soils.

But the soil of the Indus Valley in the alluvial plain of Sind has not been formed from the rock beneath it; it has been formed from material that has come from varying, and mostly great, distances.

The greater part of the preliminary weathering has been done in the Himalaya, Karakoram and Hindu Kush mountains, which have crumbled under the action of frost, heat, ice, snow and rain, the material being carried by the innumerable streams and rivers that unite into one great river near the border of Sind, some 300 miles from the sea.

The alluvial plain of Sind is the result of this river's annual floods. Sind, therefore, has no fear of the surface wastage of which Chamberlin spoke, as contributions to the soil here have been made on a far more generous scale. The highest mountains in the world have paid their annual tribute to Sind for countless years, in the form of the thin layer of silt which is deposited by the flooding Indus. The soil of Sind is, therefore, very deep compared to weathered soils. In place of the four feet of weathered soil of which Chamberlin spoke, there is as much as 40 feet, formed by waferlike sheets of mud.

Further, in contrast to stationary soils, there is not any sharp distinction between soil and subsoil. An alluvial soil, seen in the cutting of an embankment, is featureless, but it also lacks uniformity, for it is the result of a series of irregular floods. The two soils, stationary and alluvial, are quite distinct.

Here, then, there is ample opportunity for the objective faculties of humanity to exercise themselves – different soils, different treatment. If on the other hand a subjective view is to prevail, there lies a trap – different soils might receive similar treatment. Have humans avoided this trap, or have they let themselves be caught in it? Let us see.

Let us here again quote Chamberlin:

> 'Some of the soluble substances formed at the base of soils are necessary plant foods, while some are harmful; but more to the point, *all are harmful if too concentrated.*
>
> It is necessary, therefore, that enough water should pass through the forming soil, and on down to the groundwater and out through the underdrainage, so as to carry away the excess of any of these substances.
>
> An essential part of the best adjustment is thus seen to lie in *a proper apportionment of the amount of water which goes through the soil.* If this is not enough, plants will suffer from saline excess.'

I myself have been able to examine the alluvial soil in Sind. I have been able to study this soil in a cutting ten feet deep, above which the surface was soaked by irrigation. The water sank through the entire ten feet of soil, and disappeared into the earth at the foot of the cutting.

Just at the edge of the cutting, there grew a border of stunted grass and a low ericaceous plant. Farther from the edge was the irrigated crop. When the irrigation ceased, the upper layers of soil which had been moistened by the spreading of water began to dry under the influence of the sun and wind. Eventually the upper two or three feet became quite dry, and when scraped, it powdered off as fine pale sand. Below this dry surface, however, the layers down to the foot of the cutting remained moist for months after a good soaking, as I discovered when I scooped out small tunnels into its interior. The lower layers, then, have a notable capacity of storing water and, with it, soluble plant foods.

The humble desert plants – the stunted grass and ericaceous shrub – know this, for they send slender roots straight down, through the dry upper layers to the moist layers below. Some of the roots traversed the whole ten feet of the cutting and disappeared into the earth at its foot. These astonishingly long, fine roots, in places where they are numerous, look like combed hair. Quite clearly, they only rely for a short time on the upper layers of the alluvial soil for their food and water; it is upon the lower layers that they rely for their continued sustenance.

The presence of an upper dry and lower moist area after a soaking with water appears to be common to all deep, river-made soils in arid climates, such as those of Sind, Egypt, Iraq and the like. This phenomenon was described in 1906 by Professor Hilgard, as occurring in the San Joaquin Valley in California.

He found also a third dry and airy area below the moist area, caused by the soaking water pushing the air in the soil downwards through smaller and smaller channels, until it could no longer bubble up. The compressed air in this third area prevents the further descent of water, except in cracks through which it finds its way to the ground water.

This third area not only holds up the water in the middle area and prevents further loss downwards, but it also supplies that area with oxygen, which helps the microbes in it to prepare nutrients for the plants.

The arrangement, in short, is a strikingly perfect one, as one might expect, for how else could plant life be possible in these climates? It is possible because of the storage of water, nutrients and air in the voluminous middle layers, which are themselves protected against evaporation by the upper dry layers. It is, indeed, just another example of the fact that where life is found in unusual and difficult conditions, there will of necessity be such an appropriate and delicate arrangement of factors and elements that many declare they see in it the revelation of a higher intelligence.

That the Sukkur system can be acclaimed 'great' there is no question. Its construction is indisputable testimony to great technical skill and intelligence, but is the planning and execution also testimony to higher intelligence?

One can suggest that, from the point of view of the soil itself, a lack of higher intelligence so characterizes the industrial age that its existence in the planning of the Sukkur system would be exceptional.

The Sukkur Barrage in its aims is not exceptional; it has been sponsored by the same modern money-making institutions that have made such tragic blunders in the agricultural world elsewhere.

Nature's way of soaking these soils in arid countries is precisely the same as that by which she forms them; by an annual overflow of the river. When farmers originally brought in irrigation to direct the overflow to their own advantage, they did so by building embankments to enclose large areas or basins of flat land, and then sending water into them by way of channels.

During the period of the flood, water passed from the higher basins to lower basins on the way to the sea, and in each basin silt was deposited. This method is known as *basin irrigation*. Its chief exponents have been the Egyptians. The waters of the Nile were enclosed in the embanked basins for 50 days or so, some movement occurring all the time as water passed from the higher to lower basins, eventually to be drained back to the river. In that 50 days, the soil of each basin got a continuous soaking, and upon it some of the rich silt settled. The soil of each basin was cropped each year and, after the harvest, was left uncropped until the next season of flood.

If this method of irrigation is carefully considered, it will be seen that it is an adaptation by the Egyptians of the natural cycle. The water lay upon the land for the same 50 days or so of the natural flood of the Nile, and received the same deposit of mud. Throughout their long history, the Egyptians did not alter the natural cycle. It was only in the last half century that perennial irrigation began substantially to replace basin irrigation, the reason being that perennial irrigation permitted two crops in place of one. It earned, therefore, the blessing given to those who 'make two blades grow in place of one'.

This advantage of perennial irrigation is brought about by the permanent high level of a river above a dam or barrage placed in its course. It makes constant use of the artificial high level of the river, and, using the water that flows in the river all the year round, it is obviously not wasteful, but conservative.

But there is one outstanding characteristic of perennial irrigation; *it alters the age-long habit of river-made soils in arid countries*. It treats these arid soils as if they were soils dependent upon frequent rain, for by means of locks and gates, there is an application of water every ten to twenty days.

The system increases the production of the soil not just by putting more land more frequently into use; through more frequent cropping it also makes greater demands on the soil's store of plant nutrients.At the same time, it does not cater for an annual settlement of silt as does the basin method. It gets its results through the exploitation of the reserves of the alluvial plain, and not as a result of any natural function. Thereby, a challenge is issued to nature, and of course we cannot be surprised when nature retaliates.

The retaliation takes the form of an accumulation of salines in the soil. These alkaline salts lead to a deterioration of the soil and when advanced prevent the growth of crops altogether. Jacks writes in *The Rape of the Earth*:

> 'The Egyptians, during the long period in which they used basin irrigation, lived on the soil's income and won lasting security against natural hazards at the expense of progress. With the introduction of more efficient techniques into Egyptian agriculture, the soils have steadily deteriorated. Soil alkalinity has become a serious and growing menace, and cotton yields are falling. This deterioration has been due in the main to the replacement of basin irrigation with perennial irrigation.'

Basin irrigation suits the soil, and is akin to it. Perennial irrigation, on the other hand, is foreign to it, but because of the dominance of money, the perennial form was regarded as indispensable to the cotton industry with which Egypt has advanced and enriched itself.

Nevertheless, in its success, the modern economy of Egypt has played out once more the drama of money versus the soil, with money yet again the victor. But Nature will not be ignored; the very source of Egypt's life suffers, and though present generations may gain, future ones just as surely lose. Egypt's advance to modern civilization is being paid for with soil fertility.

The late Professor F. King, who became Chief of the Division of Soil Management for the US Department of Agriculture, in his book *Irrigation and Drainage* (1898), reflected upon 'the fields of the Nile kept free from alkalis for thousands of years', and upon the present increase of salts 'to so serious an extent that many acres have been abandoned', and was struck by the thought that the great irrigators of the past must at some point have experimented with so obvious a modification of basin irrigation as the perennial system. He wrote:

> 'The probabilities are that long long ago, the more 'rational' methods now being practised were tried and found inadequate or inapplicable, on account of the accumulation of alkalis which they permitted, and the old irrigators learnt to be content with a system which, although more wasteful in some ways, still kept alkalinity under control.
>
> It is a noteworthy fact that the excessive develop-ment of alkalinity in India, as well as in Egypt and California, is the result of irrigation practices that are modern in their origin and modes, and instituted by people lacking in the traditions of the ancient irrigators who had worked these lands for thousands of years before.
>
> The alkaline lands of today, in their intense form, are of modern origin, due to practices which are evidently inadmissible, and which, in all probability, were known to be so by the peoples whom our modern civilization has supplanted.'

In India, the adjacent provinces of the Punjab and Sind have both been widely developed by perennial irrigation, and both have reacted, even in a brief span of years, with increasing soil alkalinity. *The Summary of Results* (1940) by the Agricultural Department of the Punjab States states:

> 'In the Punjab vast areas of alkali soils have come into existence.'

In Sind, there have been but a few years of perennial irriga-tion, for the barrage was only opened in 1932. Nevertheless, the 1937–8 Report of the Department of Agriculture states:

> 'This constant application of irrigation water for raising crops in such intensity has created complex soil problems, the solution of which is necessary to the success of the projected agricultural progress of the province.
>
> Though precise information is not available, it is known that there are thousands of acres of *kalai* (the local name for alkaline) land where no crops would grow.
>
> Besides these large stretches, there are scattered all over the province, almost in every holding, small pieces of *kalai* land where crop either does not grow, or grows very poorly.'

Since the opening of the barrage, precise information is not available. The few investigations undertaken where it was possible to contrast pre-barrage with post-barrage conditions, show that the warning of T. Main, Director of Agriculture in 1929, that 'under perennial irrigation one must look forward to vast areas more or less infected with salt' is a prophecy very likely to be fulfilled.

Alkalinity is already the most urgent problem in Sind, and the most effective remedy that has been found, says the report, is to soak the soil with large quantities of irrigation water – up to 16–32 inches, depending upon the degree of salinity.

In other words, the most effective remedy is a temporary return to basin irrigation.

When the soil is capricious or weak and tends to deteriorate, more is involved than a mere diminution of crops; the whole life cycle deteriorates too. In reading the report, one is impressed by the great amount of disease – and not only that of the soil – that exists in the barrage area.

While it is true that no firm link between disease and soil alkalinity has been established to date, it must also be said that nowhere has the relation between the soil and the diseases found in the life cycle it supports been properly recognized or studied. The idea does not feature prominently in official agricultural reports anywhere.

Cotton is the crop to which the barrage system is particularly suited, yet in Sind, this fluffy beauty is as delicate as a Brighton invalid. Its enemies and diseases include jassids, white ants, pink and spotted boll worm, black-headed cricket, dusky cotton bug, lucerne caterpillar, red pumpkin beetle, root rot, boll rot, and red leaf. So it is officially stated:

> 'There is no doubt that the losses suffered annually by the cotton growers of Sind, on account of damage to their crops by insect pests or fungoid and bacterial diseases, are immense, and scientific research work on these pests and diseases is most urgently required.'

To compound matters, there is the effect of animals. The system was not designed for Sind's famous red cattle, as is evidenced by the fact that:

> '...since the commencement of perennial irrigation, the yield and quality of the *jowar* crop in Sind have deteriorated in many tracts.'

Jowar is a common food for cattle, and with its deterioration, there has followed a general deterioration in the quality of the stock. A heavy toll in animals is taken by such diseases as liver fluke, rinderpest, parasitic gastritis, haemorrhagic septicaemia, and so on.

Lastly, we come to the human phase. The chief disease which affects the people of Sind is malaria, concerning which the Public Health Report of 1938 states:

> 'Its incidence has increased with the inauguration of the Lloyd Barrage and Canal Construction Scheme.'

It is not possible to compile accurate statistics in rural Sind, but the prevalence of malaria is brought home to landowners by its weakening effect on the labourers, which is most evident when there is the greatest call for their labour. Some harvesting has even been abandoned altogether because of the shortage that malaria produces.

The increase of malaria occurs in the following way: the perennial system brings more water; more water brings more pools; more pools bring more mosquitoes, and the the mosquitoes spread malaria.

But this, quite possibly, is not the whole story. The Sind soils tend to be alkaline, with a pH of over 7. Low degrees of alkalinity can be neutralized by the carbonic acid which plant roots secrete, but if alkalinity increases, the water of the soil cannot hold iron and manganese to the same degree that it can when it is neutral, and *these two are the chief metals of the red matter of the blood.*

Under the barrage system, only a small investigation of pH levels has been made, but it has been found that the pH has risen from an average of between 7 to 8.5 in pre-barrage days to an average of between 8 and 9.5 after the scheme was put in place.

If this is generally true, then the plants in the post-barrage period as eaten by the Sindhis would have less of the metals that form the strength of human blood. Malaria in particular is due to parasites in the blood itself. The weaker blood favours the parasites of malaria; and so malaria thrives because of subtle changes in the life cycle, and not merely because there are more pools and more mosquitoes.

Whether this sequence is to be accepted as a further example of how we humans should be thought of as being part of a life cycle, or whether it is rejected, there can be no question that the 'growing of two blades where one grew' in Sind has ushered in a cycle of sicker soil, sicker plants, sicker animals and sicker humans. Were there a scientific measure of character and morals, it is possible that these too would be found to have deteriorated.

Jacks says that alkalinity is not as dangerous as erosion, because it can be remedied. The most effective remedy in Sind and elsewhere is the soaking of the soil. Rice growing is also effective, for in the growing of rice, the soil is covered with water and thoroughly soaked. Both processes are of the nature of basin irrigation, which in Egypt for so many centuries completely protected that wonderful land against alkalinity.

What will be the end in Sind? Will the stubbornness of nature and her dominion over all life once again force men to comprehend, or will it make their habitations barren? Is a story that is old in the East be repeated here? Sind, like Egypt, is buying its way into a civilization based on money with the fertility of its soil. Will the agricultural scientists, obedient not to the soil but to their urban masters, enable money to hold its position against the affronted land? Will they, using their fragmented methodologies, be able to go further and establish a stable and healthy life cycle in Sind, or will they fail?

For an answer we can look to the late Professor King, who possesses both a wide and wise vision in these matters:

> 'The alkali lands of today, in their intense form, are of modern origin, due to practices which are evidently inadmissible, and which, in all probability, were known to be so by the people whom our modern civilization has supplanted.'

16

Fragmentation

WHEN, in the first half of the industrial era, the demands on the soil for food and raw material became urgent, scientists set out to study the means by which the soil is able to create more life. They did not do this through the observation of nature in the forest, prairie or elsewhere, nor by a study of successful farming, past and present. Instead they did what is typical of scientists; they selected just one aspect of the question, and concentrated on it.

They studied plant nutrition - not in its entirety, but instead took a fragmented view by considering just its chemical character. They split off one part of the whole problem, and attempted, by intense study, to make the part solve the whole. In doing so, they eventually made the purely chemical aspects of plant nutrition more important than the whole.

The scientists who set out to solve this problem - in other words, to establish a scientific theory and practice of plant nutrition in the place of traditional observational knowledge and practice - were not farmers. They were chemists, just as it was chemists who were then making the factories so successful.

The three leading men in this venture were Theodore de Saussure, Justus von Liebig and J. B. Lawes.

Liebig (1803-73), with whose name the venture became chiefly connected, had already won wide recognition as one of the greatest chemists of his time. In 1832 he published, along with Wöhler, a memoir, *Researches on the Radical in Benzoic Acid*, in which he showed that the radical benzoyl could be regarded as forming an unchanging constituent of a long series of compounds. In doing so, he opened a new era of organic chemistry, and made possible almost infinite combinations of a few elements, such as those which we covered in *Chapter 14*.

Enjoying great prestige and widespread recognition, in 1838 he turned his attention to a subject that was urgent and immediate; the production of food. This was during the time known in England as the 'Hungry Forties'. In Germany it was a time of grave social unrest amongst the growing industrial population.

Liebig became drawn to the nature of food, both vegetable and animal, and he made out a very strong case for a chemical approach to it.

He rejected the traditional idea that plants derived their chief nourishment from humus formed by the decay of dead animal and vegetable matter. Instead, he taught that plants took their nitrogen and carbon from the air and eventually returned them to the air through the processes of putrefaction and fermentation. There was no loss of either carbon or nitrogen in this cycle.

But it was a different matter with the minerals that plants required, such as phosphorus, potash, soda, sulphur and lime. These came from the earth, in which they were limited and could become exhausted. Farmers had to make good the loss by returning the required amount to the soil. To affect this, they had to put themselves in the hands of the chemists.

The chemists would take some of the crops to their laboratories and burn away the organic matter, so they could analyse the minerals of the ash that was left. They would then discover whether phosphates, potash, sulphates, lime – and also nitrates, since nitrogen was not supplied speedily enough by the air – were lacking.

By mining the deficient salts or by manufacturing them in factories, they would supply those that were required by the soil. The ones particularly required were nitrogen, phosphorus and potash, as well as lime which had long been applied to fields in the form of marl or chalk. These chemicals became known by the term 'artificial' fertilizers, in contrast to natural or organic manures.

Circumstances could not have been more favourable to artificial fertilizers than at the time of their introduction. The concept of the farmer's partnership with the soil, as had been embodied in the free peasantry, had been overthrown. The old feeling for the land as something living and creative had disappeared along with the peasantry. The land had become something merely to be owned and worked for money. Large new populations needed food and other raw materials derived from plants. Fortunes were made through the ownership of land just as quickly as they were being made through owning factories. The land of England had been seized by wealthy and ambitious men, and the peasants had been evicted from their holdings and their commons. The peasants were as thoroughly subjugated by the rich as if they were a conquered people, and not fellow countrymen.

The urban areas were rapidly ceasing to have the character of country towns, and were becoming almost purely industrial. The leading industrialists had defeated the countryside. They had destroyed the rural cottage industries and had forced the young and able-bodied country folk to serve in their factories. Whether on the land or in factories and mines, the new order was ruled by rich men and their servants, the urban proletariate.

This was the condition of the rural areas at the time of

the advent of artificial fertilizers – and from the point of view of life cycles, it was bad.

On the other hand, it may be claimed that these chemical fertilizers made a big contribution to feeding the new urban populations. Apart from a few exceptional farmers, the soil had long been indifferently manured. The elements which artificial fertilizers supplied were sorely needed, and larger crops certainly resulted from their use.

Healthier and better results could have been obtained through the systematic collection of the great quantities of urban and rural waste and the manufacture of it into manure, but there were difficulties.

Firstly, the roads were bad. Even the best roads were such that sometimes even royalty could not get from Kensington Palace to Richmond because of the mud. If this was so where royalty passed, then the collection and distribution of goods needed by farms and villages was certainly not likely to be considered. The concept of efficient transport did not apply to the farming of that day.

Secondly, the making of manure from wastes requires planning and labour, and there was a lack of both in the English countryside.

Artificial fertilizers had many advantages over natural manure. They were either mined or manufactured, they were much less bulky to transport, and they were easy to spread on the fields. They were, indeed, almost *too* practical and convenient; they were easy to apply, and as they gave quick results, landowners and large farmers were satisfied.

Science thus came to the rescue and won a popoular and much-needed victory. Artificial fertilizers did great service in a period when the unprecedented rise in population and the numbers of new towns demanded intense exploitation of the soil in a country that possessed only backward agriculture.

The new fertilizers provided a partial and artificial fulfilment of the rule of return. They singled out the most important elements of plant food and replaced them, even if distant islands had to be stripped in the process. This feeding of the land was certainly superior in its results to no feeding at all. It was planned and conducted under skilled guidance. It increased yields, strengthened weakling crops in their growth, and filled in gaps caused when the introduction of motors and tractors led to losses of organic manure through the displacement of horses and oxen. In consequence, artificials came to be used in large quantities in many parts of the world.

Nevertheless, they were and still remain fragmentary; they are not a *full* return of all that is taken from the soil. Let us consider the notion that the new fertilizers have not had a whole effect, as might be assumed at first glance.

Firstly, there is something in keeping with the spirit of that exuberant time, when widespread wonder was aroused by the many triumphs of the scientific method. It is this: *the life cycle was not used as a test of artificial fertilizers.*

There is the justly famous small plot of ground of Broadbalk, Rothamsted, the experimental station founded by J. B. Lawes, where for a century wheat has been grown yearly on soil with a full complement of artificials, next to a similar plot where farmyard manure has been used. The wheat on both plots looks healthy and yields well. But all the tests of this century of experiments have been concerned only with the crop itself and its quantity. The crop has been studied as a thing in itself, by the close, careful, and fragmented watch of science. It has just been a market test – keeping to the careful and rigorous precepts of the scientific method – of the quantity of wheat yielded by a plot of land. No qualities of the wheat as a food have been tested, plus *the seed used in the plots has been imported from outside*, bringing in qualities of life cycles not belonging to the plot. So the whole century-old experiment has been without any life cycle tests, in fact it has belonged to no life cycle at all. The whole exercise has been individualized, separated, and specialized; *in nature nothing is like that.*

This fragmented approach to agriculture, as in other disciplines of science, became the standard method of research. Scientists, it is true, do test crops and foods on animals, but in a fragmented and apparently inexhaustible manner, and, except the estate planned by Lady Eve Balfour in Suffolk before the outbreak of war, there was in Britain no experimental farm on which the life cycle was the standard of test.

There seems to be, then, only one way to determine the effects that artificial fertilizers have on the life cycle, and that is to take a general view of the results of agriculture during the period in which artificial fertilizers have played such a prominent part.

Firstly, we will consider quality as measured by *taste*. F. Secrett, a farmer attending the Royal Society of Arts in 1935, spoke of taste and quality:

> 'I notice that in Covent Garden and the larger provincial markets, those stands are favoured where the produce has come from farms which have used organic manure. Although higher prices are charged for this produce, it is sold out first.'

Taste and choice are, of course, natural measures of food, but they are not scientific ones. People today so often hear statements such as 'proved scientifically' or 'measured scientifically' that they fail to realize that such variables as 'taste' are *not* measurable, and are therefore disregarded by science.

The customer who likes the look of a basket of

gooseberries and takes one to taste is a sound researcher, but he is not a scientist. Science cannot measure appearance and taste. (A great and honest scientist, Charles Darwin, once said that his work had spoiled his appreciation of music.)

But we 'ordinary' people can still judge by taste, and it has been noted by those who grow vegetables and fruits on land where full return is practised that customers often express surprise and delight when they first bite into their products. They have come to expect the bland taste of market garden produce. Anyone who has tried both foods grown from full return and those using artificial fertilizers immediately recognizes the distinction. One is inviting, the other insipid. Market gardeners themselves know it, but now that cars and vans have driven out the huge horse population which once belonged to the towns that the gardeners served, they have been left mostly helpless.

Animals, too, know that taste is a safe guide to good food. Experiments have been performed with mice in which they have been given two troughs, one filled with grain grown using biodynamic methods, which is a 'whole' method, and one with grain grown using artificial food. The mice invariably chose the first trough and finished the grain there before they went to the second. Similarly cattle, in a field equally divided into 'artificial' and 'whole return' areas, will gather and graze upon the 'whole return' area.

Nevertheless, it is surprising how few examples there are of real choice of food. The curious fact emerges that *the taste of fresh food is no longer regarded as a guide*. The great majority of modern foods are no longer expected to taste as they should by the vast majority of consumers. 'Taste', in the form of condiments, sauces, curries and so on, has to be added to them.

The next test of quality is *health*. Is food healthy when

it is produced by modern farming in which artificial fertilizers have come to play so dominant a part?

To answer this question through personal observation, one would have to go on a tour and see for oneself. We shall undertake a less laborious journey, one readily achieved by going through a textbook on modern farming and treating it as a guidebook to a country that is new to us.

As the majority of people do not know the farming world, such a book will act as a guide to what will probably seem a foreign country to the reader. I have such a book before me, written with the technical skill which one hopes to find in such books. I have read through it several times with the spirit of a traveller seeking to know what this new farming world is like, and each time I have wondered more at what I read.

There is first the soil. In this new country, one soon realizes that 'modern' soil is not a bit like the soil in nature; part of the general life cycle. It is a thing in itself, the composition of which is understood by scientists as something that they can manipulate, separate into its several parts, and dispense to farmers just as pharmacists dispense medicines. What the soil did in the past and still does where left to nature, what it did under the cultivation of past farmers, these are things of the Dark Ages that were dispelled when the 'light' of modern science came to the world.

As custodians of knowledge, the traditional farmers are not even mentioned. Previous knowledge and tradition of the land are regarded as not worth mentioning. Most modern scientists see no difference between technique and vitality.

The manipulated soil gets a number of diseases, so the modern farmer will get his soil treated by a soil scientist, just as the townsman gets himself overhauled by his local doctor. We will not discuss these numerous complaints and their treatments here.

The spontaneous power of the soil, through which it has done its job of supporting life for endless vistas of time, is lost in this new country. Even though one may know the language, one must also know the exact meanings the scientist gives to his words. One wonders what exactly he thinks of the soil. Is it a vigorous creator of life, or a cantankerous invalid? Is it the peasant's partner or the scientist's patient? It may be confusing, but let us go on with our journey.

We now enter another province, that in which the scientist is at his happiest; manipulating the breeding of plants and animals in ways first discovered by the monk Mendel. In this manner, many new varieties of life have been fashioned. Here, for example, are pigs so fat that to their ancestors they would appear as nightmares rather than pigs. They have been created for the masses of streaky bacon which the public has been taught to value. Sometimes the public taste in bacon changes, and with it the pigs are changed. These pigs are bred, fed in special ways, stalled and slaughtered, and often never see the open sky until they are taken to market. They are, of course, delicate, but they are bulky. They are tasty, too, when they reach the table, so that here public taste itself seems faulty as a guide, but we should remember that popular 'taste' is often directed not so much by consumers as it is by retailers.

Here, too, cows are specially bred for milk; they, also, see very little of the open sky during their useful life. When they become mothers and their udders fill with milk, their calves are taken from them and they are transferred to confinement in large buildings near large towns, with each animal being kept in its own stall. Antiseptic chemicals are used to cleanse the teats of microbes and then a machine is attached to the teats which sucks the milk into sterile receptacles. Extreme care is taken. Above all, the scientists test for tuberculosis, for there is probably no group of living animals so prone to a grave infection as these unnaturally kept cows are to tuberculosis.

I do have not with me, nor can I recall, the percentage of cows in Britain that have tuberculosis, but it is surprisingly large. Our guide to this new farming country, however, comes to our assistance. The best way of preventing the spread of tuberculosis in dairy herds, it says, is to test the cows and to destroy those with a positive reaction. The objection to this treatment of an invited disease is that it means a capital loss to farmers, one which they cannot afford, and so this scheme, though it might be considered to be *scientifically sound*, is not always carried out in practice.

We have looked at some of the animals in the new country. We will now consider its crops. We will look at two of the commonest staples of the human diet; wheat and potatoes.

Wheat of the best quality is required; on this we all agree, but the word *quality* in the new country has a different meaning to that which it had in the old one, when it meant a wheat that produced a health-giving and tasty loaf of bread. 'Quality' is defined these days as milling quality, or the capacity to make large loaves. Imported wheat is better in this respect than British wheat, for a two-pound British loaf is only two-thirds as big as a two-pound loaf made of imported flour. For this reason, British flour is considered weak, and imported flour strong.

The science of plant breeding is recent. It belongs to the present century, and one of its early triumphs was in making British wheat strong. This was accomplished by breeding using Mendel's principles. The *Yeoman* and other varieties show that strong wheat can be grown in Britain.

But although these wheats were strong in the baker's sense of the word, they were not strong in terms of health. They were, like the so-called soft wheats, subject to many diseases, of which those known as 'rusts' are particularly destructive. A wheat called *Ghurka* was imported from Russia because it resists rust. It was bred with British wheats, and finally there emerged *Little Joss*, which was immune to Yellow Rust. Its quality for baking was not so good as that of *Yeoman*, but its health as a plant was better.

There are many other diseases of wheat; there is, for instance, one with the charming name of 'Stinking Smut'. To avoid Stinking Smut, the seeds are soaked prior to sowing, or dusted with chemical antiseptics which are so strong that those who handle the seed are forced to protect themselves against being poisoned. Some of the poisons were too dangerous for common use, so the scientists set to work to find safer poisons.

Now those who can remember the cottage loaf, as made by hand in the countryside before these changes were begun, will recall the delicious flavour of the bread. But a delicious flavour is not such an important consideration in this new country. Consider, for example, what has been done with our second choice, the potato.

The potato is originally from Peru, the home of a great agricultural civilization. Our guide to the new country, however, tells us what a poor quality thing the potato was before being taken in hand by the scientists. Potatoes of the present day, the guidebook declares, are much superior to the small 'highly flavoured' potato of the last century. Nowadays a hygienic public, trained to associate the colour white with cleanliness, demand a potato of medium size, thin-skinned, with few eyes, and above all, it must be white. It must have a 'good' appearance; it must be a 'shop-window' potato, even though its flavour is poor compared to its yellow-fleshed, highly-flavoured and more nutritious ancestor. But this yellow colour is not 'quality'; it is considered a discoloration! Merchants will not buy such potatoes and the public will not eat them.

How is it that the public rejected flavour and nutritiousness in favour of bulk and appearance? The answer is simple. Bulk, while it means ultimately less in terms of health to the consumer, means more gain to the seller. Money wins again.

And as for appearance – is that not a second falsehood inspired by money – the eye displacing the tongue, which should serve as the sole judge in the matter of vitality? How many times have I not read and laughed at *Alice in the Looking Glass*, until, following my guidebook, I realized that I too was living in such a wrong-way-round country!

Of course, the potato has a number of diseases. We read of them in the guidebook. It is really delicate, so delicate in fact that in many countries, including Britain, certificates are issued confirming that seed potatoes are free of virus, and, in consequence, that their offspring will not die off to an extent of more than 50% because of virus diseases. There are, of course, fungus diseases as well. There is, for example, wart disease, which is so dangerous that in 1923, the government made its occurrence notifiable to the police.

So as we travel in this new country, we hear the same tale, repeated again and again. Finally the traveller must arrive at the conclusion that the modern scientific farm, and especially the experimental farm, is a mixture of forcing house and hospital. It fragments the life cycle. It is the offspring of a defect of thought, the splitting or departmentalizing of the mind, which removes the ability to see wholeness and the fact that men, animals, plants and the soil are inseparably united.

Under this fragmented view of reality, insects and other

pests – fruit flies, aphids, moths, cut worms, wireworms, leather-jackets, warble flies, maggot flies, and the rest – have assumed a dominant role which they have never enjoyed in nature.

Fortunately, as a reaction to these misfortunes that afflict the new farming, concern for the balance of nature is once again emerging, under the term *ecology*. How different is the approach of ecologists to that of scientific money-farms! Here is some evidence, taken from the *Biodynamic Agricultural News Sheet* of April 1938:

> 'We have found it possible to prevent plants from suffering damage from insects simply by taking suitable biological measures, and without taking steps to kill them.
>
> In vegetable culture, we are mostly concerned with plants whose flowering impulse is held back – for example all kinds of cabbage, carrots, radish, chicory, leek, celery, beetroot, turnip, etc; or we are dealing with plants whose flowers are not very prominent, such as beans, tomatoes, and similar plants.
>
> A close study of the relationships in nature makes it clear that the insect and plant worlds are complementary to and dependent upon one another, and moreover that certain insects and certain plants are sympathetic to each other. Vegetables enable insects to develop their larvae and flowers offer food to countless fully developed insects. Also, there are many small creatures which prey on each other, such as spiders, ichneumon fly, ladybirds, etc.
>
> If we provide as large a variety of insects as possible with the means of living, most of them will live harmoniously together, and the harm done by any particular insect will be practically negligible. That is why it is so important to have flowering plants near vegetables. The aromatic

> herbs are especially valuable for this purpose, for example borage, lavender, hyssop, sage, thyme, marjoram, dill and fennel.
>
> At first the grubs of the cabbage fly were very destructive. Now we do not mind them at all. If on warm days at the end of April or May the fly lays her eggs on the cabbage plant, the red mites find them and suck the eggs before the larvae emerge. There are many such compensatory adjustments; for example the sandfly drags many caterpillars away to bury them for its larvae. In spring, ants seek among grass and plants for the larvae of the daddy long legs and kill them. Even the wireworm made itself useful by preferring predigested plant material, and destroying the larvae of the cabbage fly.
>
> Not only do insects thus achieve balance amongst themselves, but toads, frogs, moles, shrew-mice and lizards also take part in this adjustment.'

Men, animals, plants and the soil, then, are balanced and united. That is the clear conclusion of the *Biodynamic Agriculatural News Sheet*. The ignoring of this fact is the cause of all the strangeness in the new country through which we have travelled. It is also raises questions about one particular aspect of the new world; artificial fertilizers.

Humans separate and fragment. They separate the science of chemistry from nature. Then a common consequence of bad thinking happens. The new specialists, the chemists, look at nature's vital processes from a purely chemical point of view, and claim to be guides and masters. Being scientists, they get the support of other scientists, and that of the 'scientific method' – the adherence to experiments which can be repeated.

Thereby they fragment, isolate and simplify questions, and make them readily 'comprehensible' and 'controlled'.

With the use of artificial fertilizers, they reduce the feeding of the soil to a purely chemical process, and by ignoring the secrecy, delicacy and variety of nature's own methods, they also limit their own chemical science. They fragment and simplify it to just a few minerals; nitrogen, phosphorus, potassium and calcium. But there are more minerals in the soil than are dreamt of in their philosophy; for example, those that accompany the growth of sugar beet by the sea constitute, as Dr Pfeiffer describes it in *Biodynamic Farming*, a small pharmacy of sodium, lithium, manganese, titanium, vanadium, strontium, caesium, copper, rubidium, some of which are as rare as their names are exotic.

As to these elements, both rare and common; do we know what tone and quality we miss if we lack our share of them? We already know from *Chapter 14* that grave defects can arise from small omissions, but the mind need not get confused contemplating the possibilities. We have just read of the wonderful balance of nature's adjustment between insects, flowers and other small forms of life.

Will not nature establish the same balance within us, if only she is allowed her way, and we follow, and do not disrupt, her life cycles? Will not each element harmonize with the others and so bring about a state of health? And how can we expect health, when harmony is missing in the soil? This is why the effect of artificial fertilizers is so important.

I could give other answers and show how, when the whole of Nature rather than its parts is followed, health must necessarily accompany it, but I will not do so, as I have already made this the subject of my book *The Wheel of Health* (1938).

I will end this chapter not with minerals, but humans.

Our agriculture is wrongly based. It is a system largely

directed at curing evils which it itself is responsible for. It is the wisdom of the country and the traditional farmers we need now; that of those who have built up long-lasting agriculture and whose wisdom lies in tradition. They have fashioned it through physical work and close and immediate observation – through the personal intimacy with nature which we have come to associate with the poet.

And, in fact, peasant life *is* poetic, and it is so precisely because of this intimacy. The music, dance and art of peasants are the creative expression of their lives, and as such are characteristic of their environments, including the land on which they live.

Nothing collective or traditional, as peasant life is, originates from people separated from the soil, as are townfolk. The poems and essays that played a notable part in the country life of the Chinese, the Tibetan art which finds its way into every home, the sylvan setting of Japanese villages, of the Balinese and Burmese, the vocal harmony of Swiss peasants returning from their fields, the reproduction of floral beauty and colour in festive dress of so many countries – these are the product of the poet that lies in every peasant's heart. It is this intimacy that inspires creativity in the poet, as the Greeks recognized in their choice of word for poet, namely, a 'maker' or creator, and which Dante voiced in the *Divine Comedy*, when he wrote that the poet was not the disciple of the imagination, but rather one who knows the secrets of nature.

It is this intimacy which reveals to the peasants and country folk the complete, interdependent character of all the varying forms of life, and the health, goodness and beauty which come from it. Its all-pervading quality is something known by being seen, felt, lived with and realized, and not just told or read. It constitutes that mystical unity with which all the most meditative religious thought and all the most sublime art have concerned themselves. The most famous temples of the world, the noblest poems, the loveliest pictures, the most transcendent music; all have acclaimed it. They are all works of balance and beauty, created by the realizations of artists who know the secrets and appreciate the mysteries of nature.

It is to this company that the knowledge and arts of real rural life belong. Great or humble, they share the same origin.

Modern urban civilization, on the other hand, divorced from the creative power of the soil, has forsaken this great heritage. In its place there has spread a nihilism that year by year has been destroying art, truth and beauty, and, at the same time, to an immeasurable degree, the soil itself – a process which threatens to be consummated in a holocaust of humanity and nations.

17

The East and West Indies

THE DUTCH BEGAN THEIR CAREER as empire-makers in the East almost at the same time as the British, who had a head start of only ten years.

Both peoples came to the East to trade, and both became imperialists almost accidentally, as result of their being traders. There was a notable difference, however, between the Dutch and the British, even though they were close neighbours in Europe.

The difference lay in their attitudes to their own soils, and this difference they carried with them to their eastern possessions. Consequently the Dutch colonial governments left the cultivation of the acquired lands to the native farmers, without any interference other than requiring the payment of taxes in kind, through which they acquired the tropical products they needed for sale in the markets of Europe. Everything concerning native agriculture, subsistence farming, the village system, and native rule was left undisturbed.

It was a method of rule similar to that which the British now follow in some of their West African colonies, and which the colonial statesman Lord Lugard described as the *dual mandate.*

In the latter part of the nineteenth century, the Dutch government of the islands gave up the role of trader. By then it had become so firmly convinced of the value of the Javanese and other peasantries that it protected them through the absolute prohibition of the sale and purchase of land.

A Mr. Boys of the Bengal Civil Service visited Java in 1892. He summed up the preservation of the Javanese peasantry in this remarkable passage:

> 'The Javans have escaped the fatal gift of proprietary right, which has been the ruin of so many tens of thousands of our peasantry in India, and with which, while striving to bless, we have so effectually cursed the soil of India. It is not too much to say that the many benefits which would have been conferred on Java by the substitution of the English for the Dutch rule, were not too high a price to escape from the many evils of the unrestrained power to alienate private property.
>
> Under their present government, the Javans should, according to our English ideas, be the most miserable people. That they are not so, but that, on the contrary, they are the most prosperous of Oriental peasantry, is mainly due to one cause – the inability of the Javan to raise a single florin on the security of his fields, and the protection thus gained against the moneylender and himself.
>
> Nature is bountiful in Java, and the abundant fertility of the soil enables the Javan to stand up against many ills to which he is subject; but were Nature's fecundity doubled, were she able to pour her gifts as from a cornucopia into his lap, nothing would be able to save him from the moneylender and from danger of eviction from his fields and his home, if he were able to pledge either one as security for an advance.'

The Javans carry out a very skilled agriculture and have got erosion under as complete control as has been achieved anywhere in the world, according to Jacks. He continues:

> 'The Dutch government in Java has carefully preserved and encouraged native agriculture,

> and the same principles are applied to European-controlled estates. There are no social barriers between European and natives in Java. The primary object of agriculture is to feed the people; the food supply of the community must be maintained on a permanent and secure basis before rubber, tobacco, coffee, etc., can be produced for export.'

Here, everything has been favourable to the soil. The traditional cultivation and anti-erosion measures of the Javanese are excellent; the early Dutch government was able to get what it required through taxes in kind, without other interference. The government came to value the peasants' skill so highly that it did everything to support it. In doing so, the Dutch encouraged a sane and sensible agriculture, *the primary object of which was to feed the people.*

Through the relative absence of social barriers, both peoples, Dutch and Javans, were able to base themselves on the soil. As a result, the soil is valued by both races.

In Java, writes Jacks, they

> 'give every acre of land a national value that is out of all proportion to its money-making power'.

This high valuation of the soil was held by the indigenous population, but one can well understand how highly it was also appreciated by such a people as the Dutch. When they came to Java, the Dutch were the best cultivators in Europe. But they were something in addition.

Of all peoples in Europe, they had waged the greatest and most unceasing fight for the preservation of the soil; they, above all people in Europe, had given land a national value. They had for centuries won land from the sea and flood and guarded it by dykes, which they maintained perpetually. In their knowledge of the use of water and drainage, in the rotation of crops, in the use of clover, in all the arts of cultivation, the Dutch in the Indies were more advanced than their British contemporaries.

The improvements in agriculture in England which were initiated at the time of Elizabeth were due largely to Dutch and Flemish influence and infiltration. It was Charles the First who brought Dutch experts in dykes and drainage to make his estate of the Isle of Axholme the best worked in England. Of all the Western peoples, therefore, who could have colonised Java and its sister islands – if such a thing was destined to occur – none could have been better chosen or suited than these skilled and soil-revering farmers of north-western Europe.

So, in this respect, the Dutch East Indian islands had the advantage over the British West Indian Islands. Of the quality of the Eastern cultivators themselves, there is no better account than that of C. Wallace in his book *The Malay Archipelago* – firstly, because he was a skilled observer, and secondly, because he visited the islands 70 years ago and, in the island to be described, saw its agriculture as something unique.

The island is that of Lombock, separated from Java by the island of Bali. It has at present some 600,000 inhabitants. Its capital is Mataram. Wallace wrote:

> 'Soon after passing Mataram, the country began to rise gradually in gentle undulations, swelling occasionally into low hills towards the two mountainous tracts in the northern and southern parts of the island.
>
> It was here that I saw one of the most wonderful systems of cultivation in the world, equalling all that is related to Chinese industry, and as far as I know surpassing in the labour bestowed upon it any tract of equal size in the most civilized countries of Europe. I rode through this strange garden utterly amazed, hardly able to accept the

fact that in this remote and little-known island, from which all Europeans except a few traders are jealously excluded, many hundreds of square miles of irregularly undulating country have been so skilfully terraced and levelled, and so permeated by artificial channels, that every portion of it can be irrigated and dried at pleasure.

Depending on the slope of the ground, each terraced plot consists in some places of many acres, and in others of a few square yards. We saw them in every state of cultivation; some in stubble, some being ploughed, some with rice crops in various stages of growth. There were luxuriant patches of tobacco, cucumbers, sweet potatoes, yams, beans or Indian corn. In some places the ditches were dry, in others streams crossed our road and were distributed over lands about to be sown or planted. The banks which bordered every terrace rose regularly above each other, sometimes rounding an abrupt knoll and looking like a fortification, or sweeping round some deep hollow and forming structures resembling the seats of an amphitheatre. Every brook and rivulet had been diverted from its bed, and instead of flowing along the lowest ground were to be found crossing our road half-way up an ascent, yet bordered by ancient trees and moss-grown stones so as to have all the appearance of a natural channel, and bearing testimony to the remote period at which the work has been done.

As we advanced further into the country, the scene was diversified by abrupt rocky hills, steep ravines, and clumps of bamboos and palm trees near houses and villages; while in the distance we could see the fine range of mountains of which Lombock peak, 8,000 feet

> high, is the culminating point. It formed a fitting background to a view scarcely to be surpassed in both human interest or picturesque beauty.'

This naturalist and observer, it will be noted, is struck by 'one of the most wonderful systems of cultivation in the world', 'in a remote and little-known island, from which all Europeans except a few traders are jealously excluded'; a cultivation due to assiduous labour; a skilled and complete use of water including every rivulet and brook; the roads themselves are made subject to the water channels; the levelling of every plot of land, large and small, so that the water could be equally distributed; and he is able to end his description with a true and happy association of 'picturesque beauty' and 'human interest'.

The British in the West Indies met with no great indigenous cultivators like those of Java, Bali and Lombock. The British invaders of the West Indian Islands were buccaneers and adventurers, who at great risk and for their own personal gain set out upon the high seas to dispute the Pope's fiat that the New World, both known and unknown, belonged to the Spanish King. They and the French took many of the islands from the Spaniards and made them their own.

The early history of Jamaica is a typical example. The island was discovered by Columbus in 1494, and taken over by the Spanish. With a criminality towards indigenous peoples which seems to have been peculiarly their own, the Spanish annihilated Jamaica's gentle and peaceful inhabitants.

When the British took the island from them, the total of Spanish masters and their slaves did not exceed 3,000. So the British, some of whom had left their homeland for love of adventure, and others who had left for fear of the law, now found fortune before them. In 1672 the Royal

African Company was formed, and Jamaica became one of the busiest slave markets in the New World. The cultivation of sugar was introduced. Pepper, coffee, cocoa, ginger and indigo, products sent from Java to Holland by the Dutch, were now sent from Jamaica to Britain. When slavery was abolished in 1838, the prosperity of Jamaica was at its height.

In Java, as we have seen, the management of trade was undertaken by the government itself, and the work was carried out by the indigenous people using their traditional methods. In Jamaica, the work was carried out by black slaves owned by planters, whose object was to enrich themselves, with or without benefit to any partners in Britain. The personal motivations of the buccaneer remained with them; namely, that of using their property primarily for their own personal advantage. Both the soil and labour were their slaves.

Slavery, while it endured, carried with it the obligation of the planter to feed and house his slaves upon the estate. So, the first function of the soil – to feed the people who work upon it, was fulfilled. There was a direct relationship between the workers and the soil upon which they worked. The planters, as slave-owners, had also a direct relation with their workers, a position that in the case of many of them who were moderate and possessed some humanity, amounted to a paternal form of chieftainship. They stood in a parental relationship, such as is described in a recent best-seller, the novel *Gone with the Wind* – a relationship which has, in its better forms, constituted the main human binding power of the British Empire, and which has acted as a restraint of the power of money.

W. Macmillan, in his *Warning from the West Indies*, writes:

> 'Slavery was not, as some maintain, wholly evil in its effect on the slave-owner's character. It not only fostered a proprietary sense of responsibility; slaves made possible a spacious leisure... Many fine planters in the West Indies and the Southern States, like some Cape farmers, have a delicacy of culture associated only with the choicest traditions of old Europe.'

Such culture could make good masters. The prosperity of the planters overflowed in generosity towards their dependants. It could even be argued that, in terms of happiness, the West Indies were well off in the eighteenth century.

Great change was brought about by the emancipation of the slaves in 1838. This emancipation was an act of liberalism, but there is something greater than liberalism, and that the wisdom that comes from the soil. And in the light of the soil, this emancipation was superficial and unreal... agreed, it was a liberation of slaves from servitude, but it was also the release of the planter from his responsibilities.

The dominant factor in an agricultural island is not slavery or freedom, but the subsistence farming and craftsmanship of the peasants, their families and their kinsfolk.

There are only two relationships of agricultural workers to the soil.

The first is that of *slavery*, when they are assured of their subsistence from the soil and are valued by their owners and kept in health and happiness, because the estates are then well-worked and conserved, and the owners give to the estates the quality of a home. A certain easy and ready acceptance of life, with the rich flavour of a landed aristocracy, comes into being, and places the whole art of life on a plane which stands above that of land as a mere agency for the market and for profit. The buccaneer becomes a gentleman and the slave a devotee.

Nevertheless, money becomes paramount when the freedom and wealth of the landed gentry becomes shackled by the middlemen of the town. Money takes command.

Moreover, the means of highly skilled, soil-conserving agriculture are not acquired by slaves, because they do not have the right to own property. The meticulous care of the soil which is required for truly sustainable agriculture seems to be the possession of only the second form of human relationship with agriculture; that of the *peasant family.* The self-dependence of free peasants produces qualities, understanding and practices that are absolutely superior to those of slaves.

The emancipation of slaves converts them, supposedly, into free individual labourers. This change was experienced in the West Indian arena in 1838. It has existed now for a hundred years and recently celebrated its centenary – logically enough as we shall see – with riots and revolts.

Harold Stannard, writing in *The Times* in 1938, described the dwellings of the humble agrarians in Jamaica:

> 'The first time I saw one of these hovels, I could hardly believe that it was intended for human habitation. Strands of dried bamboo are woven round a framework of stakes and the 'room' thus formed is covered with palm thatch. There is no furniture, except sacking on the earth and some sort of table for the oil stove... urban conditions are, if anything, worse.'

Royal Commissioners declared the slums of Port of Spain, Trinidad, to be 'indescribable in their lack of elementary needs of decency'. Conditions of labour sometimes find the Commissioners almost wordless: 'It would be hardly possible to find terms strong enough' to express their disapproval. Such statements are but part of the general chorus which accompanies the imperial achievements of the time and which finds full expression in Royal Commission reports on labour in India, Basutoland and elsewhere.

Here is Stannard's statement with regard to nutrition and subsistence, the primary test of the relationship with the soil:

> 'Under the stimulus of a circular dispatch from the Colonial Office, inquiries have been conducted in the islands and have yielded disquieting results. Even to a non-medical eye the frequency of bad teeth among a population whose diet could and should contain a large proportion of fresh fruit and vegetables gives cause for misgiving.
>
> Indeed, it is not necessary to look into the islanders' mouths; it is enough to glance inside the shops where they buy their food. Every Chinese-kept store exhibits, from floor to ceiling, shelf after shelf of tinned goods. These superbly productive islands, living mostly by the export of food, cannot feed themselves. It is estimated that Trinidad imports four-fifths of what it eats.'

This, then, is the condition of the islands which Britain cherishes as the oldest of her colonies. *Britain has never adopted, as an unalterable principle, the right of the people to support from their soil.*

Britain brought Africans as slaves to the islands. With the ascendancy of money, the right of slaves to rely for subsistence on the soil has been taken from them, and under the cover of apparent freedom, the British have made the slaves' condition even more oppressive than it was in the past.

The words of Pope Leo XIII:

> 'Every man has by nature the right to possess property as his own. Hence man should possess not only the fruits of the earth, but also the very soil...'

...do not apply. They apply neither to the soil, nor to the fruits of this very fertile island. The power of money yet again emerges as the enemy of the people's source of life.

We continue with Stannard's words:

> 'Only by a reversal of the policy which prefers money crops to food crops can the native labourer be assured of the conditions which make a civilized life possible. Apparently the evil has increased in recent years. The Barbadian report is definite on this point:
>
> > 'In the old days, plantation proprietors planted a fairly large acreage in food crops, some of which were sold to labourers at preferential rates. But in recent years, the cultivation of food crops has been curtailed to the extent that the price of locally grown vegetables is so high as to be beyond the modest means of the labourer.
> >
> > The absence of fresh vegetables and proteins in the diet of the labourer is having a deleterious effect on his health and physique. In short, the modern methods, which have tended to divorce the field from the sugar factory and make of them distinct and separate entities of plantation economy, have worked to the detriment of the field labourer.' '

The quoted Barbadian report uses soothing language such as 'we gather', 'in short', 'have tended', 'the detriment'... to obscure the stark nature of reality. Stannard, however, is in no doubt about the need for the reversal of the policy as regards money and food crops. He writes:

> 'In the Dutch East Indies, land sufficient to meet the needs of the whole population is earmarked for food crops before any money crops are allowed to be grown.'

Therein is the difference between the British West Indies and the Dutch East Indies. The West Indies now exhibit all the signs which in Britain's towns and cities have come to be described by the slogan of 'scarcity amidst plenty'. Although there are plentiful soils in a plenteous climate, the British have nevertheless inflicted extreme poverty and malnutrition on the mass of the people.

From cold, of course, the people here cannot suffer, and so their hovels do not need to have the protective character of northern homes. But, with this sole advantage over the British to whom they belong, the people of the West Indies seem to be as far from plenty and as near destitution as a people can be.

Macmillan, quoting his personal investigations, states that the spending power of the average citizen is so low that it is scarcely above that found in one of our more recent colonies, Nyasaland.

But in many ways, the islanders can be considered advanced. The Barbadians, so many of whom cannot afford fresh vegetables, apparently take pride in calling their island 'Little England', since, though it is smaller than the Isle of Man, it supports an Established Church, two Chambers, a Court of Grand Sessions, eleven Parish Vestries for local government, and probably the best education system in the West Indies. But not enough vegetables.

Agriculturally, the Barbadians are careful and skilled cultivators of cane. Their fields are clean and well-tilled, and there is

> '...a serious effort to find and make work for as many hands as possible – for a dense population

> of more than 1,000 to the square mile – through intensive island-wide cultivation of sugar cane.'

Out of 176,000 inhabitants, some 18,000 are said to be small holders of a total of 14,000 acres; so 77 per cent of the land owners have less than an acre. According to Macmillan, the land which the peasants get for themselves is:

> '...only the poorest soil. Often it is the land of some estate ruined by its European owner's bad or indifferent cultivation.
>
> Barbados, however, despite facing a most serious population problem, is in fact dead set *against* the peasant solution. Peasants, it is held, have failed to maintain the output which has so far kept the island going, and so long as cane is the only industry, nothing but the highest possible output will suffice.
>
> The peasants, however, have had their chance only on poorer soil, without organization or even sympathetic direction. The Barbadians, moreover, have no experience as peasants, in fact they have little tradition other than that of supervised plantation labour. Although they are intelligent workers under direction, they would not be at their best as individual cultivators. In the long run, the only alternatives for Barbados would seem to be great industries absorbing much labour, and making the island more like one town – or else a steady flow of emigration.'

Barbados and Antigua lack the range of mountains that most other islands in the West Indies possess. They have, therefore, been given over to monoculture – that of sugar – and, as Macmillan states, the workers have no experience as peasants, and little tradition other than that of supervised plantation work.

The other islands, in this, have great advantages over them. Let us take Jamaica, the largest island of the British

West Indies with its 4,550 square miles and nearly a million inhabitants, as an example of an island with a central range of mountains.

Macmillan writes:

> 'The Jamaican peasant tradition is due not to any special aptitude of the slaves imported there, but rather to the fortunate juxtaposition of ample valleys and less accessible but still fertile and attractively habitable hill country; this accident gradually led the estate owners, as seldom elsewhere, to allow some of their slaves to grow their own food supplies.
>
> Thus a strong agricultural tradition was established and has persisted. After emancipation, many former slaves became independent cultivators and the Jamaicans, though they may be less disciplined, to some extent escaped the routine work that is characteristic of the other sugar islands.'

It has been said that there are no less than 150,000 smallholders. Macmillan, however, doubts this and declares that:

> '...peasant lots are now obviously too few and too small to provide an adequate living for any sufficient number of Jamaica's million inhabitants.'

A few, though, are successful and have saved money from their farming. They do not use their money to improve their land, but instead buy up the land of their weaker brethren, who then become their tenants. They are not, then, careful partners of the soil. The values under which they live are those of private property, individualism and the survival of the fittest, and do not include the controlled and sustained maintenance of a well-farmed soil.

The successful peasants imitate the white landowners.

Frequently they overreach themselves by taking too much land, while the white landowners, for their part, continue to hold their partially used estates in the hope that fortune will change and bring a better market. A further consequence of an ill-founded agriculture then appears:

> 'Control, if not ownership, passes into the hands of banks or business firms... we see the almost invisible change of control from private landlords to outside mortgagors.'

So peasant ownership as a policy languishes in the islands of the West Indies, just as it does in England. Macmillan writes:

> 'In enlightened circles of widely different views, the approved policy – insofar as there *is* a policy – is to offer opportunities to rise to peasant status to as many as possible of this heterogeneous mass of small tenants – a few of whom originally started the trend without any help.
>
> Official encouragement has been strongly given by a few individuals – notably by Sir Henry Norman, ex-Governor and head of the West Indian Commission of 1897, and following him, by Sir Sydney Olivier during his governorship of Jamaica; but it is still only an aim, not an achievement.
>
> Even the aim has usually been hesitant. It is not clear whether peasants are to be relied on for the main agricultural production of the country, or whether peasant-ownership is only a means of relieving unemployment.'

The final result of ill-founded agriculture now shows itself in these naturally luxuriant islands; erosion of the soil. Whyte writes:

> 'Different factors are connected with the erosion which is occurring in many of the islands of the West Indies.
>
> In Jamaica, tenant farmers have practised shifting cultivation – paying rent for, say, one acre but burning and destroying forest over a much larger area. In addition, accessible areas of forest have been heavily over-exploited, and there are now insufficient forest reserves. In the plantation districts all land fit for cropping has been cleared, and in addition, excessively steep slopes have been disposed of to petty settlers, for the production of foodstuffs...
>
> Deforestation has also been excessive on some of the Windward and Leeward Islands. For example, a critical stage has been reached on the island of St. Vincent.
>
> In Trinidad, clearing of forests and shifting cultivation have caused serious denudation, erosion and severe flooding in the Maracas Valley and the Caroni plain.'

Particularly valuable is the result of a survey of the United States island of Puerto Rico. It was found that

> 'there is slight erosion on 19 per cent of the island, mostly on cultivated parts of the coastal plains and alluvial valleys or on gently rolling pasture lands; moderate erosion was found on 29 per cent, and severe erosion on about 39 per cent on the area. Most of the severe erosion occurs in the rough mountainous interior. Sheet erosion is the most common type, with gullies occurring on a little less than 22 per cent of the area.'

The same author elsewhere in his book writes of two British islands, scarcely less fertile than Puerto Rico and Jamaica, where a similar wanton disregard of the soil

> '...threatens to leave the country like an emaciated skeleton'.

It seems that the haunting vision of the South Pacific now reaches the lovely islands of the Caribbean.

18

The German Colonies: The Mandates

THE GERMANS were the last Europeans to colonize. They were also the people most imbued with faith in modern science, and this led them, with a clear conscience, to apply their ideas regarding the rights of the fittest and strongest to their extreme conclusion. Armed with this faith, they conquered three territories in Africa: South-west Africa, the Cameroons, and Tanganyika.

The Germans date their colonial empire from 1884, when Lüderitz hoisted the German flag at Angra Pequena, a port of South-west Africa. Nachtigal did the same at Duala, a port of the Cameroons, and Karl Peters and his companions landed at Zanzibar. So began the German exploitation of Africa.

Many countries had preceded them. Portugal, Britain, France and Belgium, in their exploitation of their new territories, had not always refrained from cruelty. One of them, Belgium, under the influence of its king, was in the 1890s to give an example of cruelty on such a large scale, and so pitiless that, when knowledge of it became public, it caused a wave of horror through the United States, Britain, France, and Belgium itself. The period of wholesale harsh treatment of natives had come to an end as far as public sentiment in Western Europe and America were concerned.

With this establishment of the three colonies in the year 1884, the German version of the policy of exploitation began.

In South-west Africa, a dry land and chiefly of agricultural value because of its pasture land, Paul von Rohrbach defined German policy in the *Deutsche Kolonialwirtschaft*, as quoted by G.Steer in *Judgement on German Africa* (1939):

> 'The decision to colonize South-west Africa meant nothing less than this: that the native tribes would have to give up their lands on which they had previously grazed their stock, in order that the white men should have the land for foraging their own.'

The Hereros and Hottentots were the indigenous peoples most affected by this appropriation. It began with the harsh oppression of both peoples, particularly of the prouder and more warlike of the two, the Hereros. One of their chieftains described the German methods, again quoted in Steer's book, which is my guide in this chapter:

> 'Our people were robbed and deceived right and left by German traders. Their cattle were taken by force. They were flogged and ill-treated and got no redress. In fact, the German police assisted the traders instead of protecting us.
>
> Very often, one man's cattle were taken to pay another's debt. If we objected and tried to resist, the police would be sent for and, what with floggings and threats of shooting, it was useless for our poor people to resist. If the traders had been fair and reasonable, like the old English traders, we would never have complained. But this was not trading at all. It was only theft and robbery.'

The Hereros rebelled in 1904, and fought according to their savage code, seeking revenge. They were defeated and, to finish the work, General von Trotha issued an order for total extermination; the *Vernichtungs-Befehl*:

> 'I, the great general of the German soldiers, send this letter to the Herero nation. The Hereros are no longer German subjects. They have murdered and robbed, they have cut off the ears and noses and privy parts of wounded soldiers, and they are now too cowardly to fight...
>
> The Herero nation must now leave the country. If they do not do so, I will compel them with the big tube. Within the German frontier every Herero, with or without a rifle, with or without cattle, will be shot. I will not take over any more women and children, but I will either drive them back to your people or have them fired on.
>
> These are my words to the nation of the Hereros.
>
> *–The great General of the Mighty Emperor, von Trotha.*'

By the end of 1905, German extermination had reduced the Herero people from 90,000 to 15,000. In October 1904, the Hottentots also rebelled, and suffered a similar fate.

As to the human result of these policies, Leutwein, the German historian of the South-west, declared:

> 'At the cost of several hundred million marks and several thousand German soldiers we have, of the three business assets of the Protectorate – mining, farming and native labour – destroyed the second entirely, and the last by two-thirds.'

Before the Germans were themselves conquered in the first World War, the condition of the natives is summed up by Steer:

> 'The native was a State serf, guilty of serf-like offences. Out of 4,356 convictions against natives in the Protectorate between 1 January 1913 and 31 March 1914, 3,167 were for desertion, negligence, vagrancy, disobedience,

> insolence, laziness and contravention of the Pass laws; crimes not of man against man, but of a slave against his owner.'

This did not include the punishments of *Väterliche Züchtigung*, or paternal punishment, allowed to the German master over their serfs, which led Governor Seitz, in order to avoid a further native revolt, to threaten in 1912 to withdraw labour supplies from those –

> '...who rage in mad brutality against the native, and consider their white skin a charter of indemnity from punishment for the most brutal crimes.'

After the first World War, South-west Africa was allotted as a Mandated Territory to the Union Government of South Africa.

In the Cameroons, the Germans adopted the same policy, but it did not lead to any rebellion and annihilation such as that of the Hereros. The policy here was to hand over both the land and the natives, as and when required, to the great German commercial companies.

Governor Jasko von Puttkamer was a leading advocate of these companies, and with typical German efficiency, he turned over the colonial government to their interests.

'Administrative recruitment' meant that the natives were used as the planters and traders needed them. At first rubber was the chief source of wealth, but when Puttkamer in 1895 saw the coffee, cocoa and banana plantations on the neighbouring Spanish island of Fernando Po, he initiated estates for these products upon the lower lands of the lofty peak of Kamerun. Concessions were given by Puttkamer to German companies, and plantations were established on land taken from the Africans, who were induced to work for the planters by being left sufficient plots on which to grow their own food.

The demand for porterage to carry the products of the new plantations to the coastal ports increased. Men were taken from their farms and families by 'administrative recruitment', to carry loads on ceaseless journeys, the police acting, as in South-west Africa, on behalf of the planters and in no way protecting the natives. For the benefit of the plantation owners, 10 per cent of the population became serfs.

Puttkamer's exactions, financial machinations and private life were rooted out by the Social Democrats in Germany. He was disgraced and dismissed in 1907, but so wealthy and influential had the planters become that his policy continued, so as to keep the labour market well supplied, despite its depletion from disease and hardship.

A better spirit prevailed in – or was forced upon – the German Colonial Ministry. A German medical service was organized to check the loss of labour due to diseases, the most feared of which was sleeping sickness. Some official attention was also paid to education. 1,000 children were to be found in government schools in 1914 and there were 40,000 in German and American missionary schools. There were reforms, but they were only minor. As far as they went, they had some benefits for the natives, but nevertheless 'administrative recruitment' remained.

The natives were still the serfs of their masters, and discipline was enforced by severe punishments. Planters were accused of the 'physical and moral annihilation' of the native, and it was not until 1914 that the Colonial Minister, Dr Solf, was able to announce a doctrine new to the Germans:

> 'The colonies will prosper with the natives and for the natives, not in spite of them and against them.'

But it was too late for this new policy to be tested on its own merits. After the first World War, the German Cameroons were divided into two. The greater share was given to the French, the lesser to the British. The French now ruled a population of 2,400,000, and the British 800,000.

The conquest of the third colony, Tanganyika, was due to Dr Karl Peters, of whom Steer writes:

> 'Of all the German pioneers, 'Hangman Peters' was the most unprincipled and bloody. I have not written of his cruelties, because I do not regard him as typical of the old German colonists; none but Trotha was as foul as the merciless doctor.
>
> But evidently the Nazis of 1934 held him to be typical – nay more, a prototype. Their propaganda has heaped praise on him in the five years since they gummed a memorial to him on their envelopes...' (His portrait appears on a stamp which celebrated Germany's Colonial jubilee in 1934, and is placed as the frontispiece of Steer's book.)

In 1888, Peters acquired Tanganyika by what Steer calls:

> 'a novel piece of international theft, to which all civilized powers were parties.'

For ten years, Peters exercised his power sadistically. Finally, his hangings and shootings of natives and the flogging of his concubines became known to the Social Democrats of the Reichstag. In 1897, he was brought to trial before the German Colonial Disciplinary Court. He was dismissed from the governorship and took refuge in England.

Nevertheless, the hatred which he had aroused amongst the natives did not subside. Too many German bullies remained behind; too many native chiefs had been robbed. One called Mkwawa rebelled, and was defeated. His German conquerors cut off his head and sent it as a trophy to Berlin. A special clause was inserted in the Treaty of Versailles which ordered its return to his tribe.

A more serious rebellion, that of the Maji-Maji, the combined tribes of the south, raged for two years. The Germans, failing to overcome the Africans in the field, destroyed their villages and crops by fire. Thousands of Africans died of starvation and the entire south of Tanganyika was devastated. Money became frightened and so, in 1907, Dr Dernburg, an able businessman and banker who had been given the new appointment of first Secretary of the Colonies, left for East Africa to institute reforms and to endeavour to turn the hatred of the Africans into their natural tolerance, if not affection. Shortly after his arrival, he announced:

> 'I saw too many whips in the hands and on the tables of the planters and colonizers.'

He attempted to permit the natives to be free producers, as well as to free those who were in German employment, limiting forced labour to public works and paying the natives for their work. The planters were bitterly hostile to these reforms, and succeeded in forcing Dernburg's resignation. The seizure of land, forced labour, floggings and imprisonment continued until the first World War.

After the war, Tanganyika became a British possession. The Permanent Mandates Commission of the League of Nations brought a redemptive spirit to governance. Under the first clause of their charter, they were to be *the trustees of the material and moral well-being of the natives.* Each Mandatory Power had to present an annual report to the Commission. Reports and comments were made public. The old-time secrecy which had been used to hide offences from public scrutiny was made impossible.

Steer, in his chapter titled *Mandates' Work*, gives an account of the improvement of the lot of the indigenous population of the former German territories. Of South-west Africa, he states:

> 'The amount of land in the hands of the natives has multiplied by ten since the days of the Germans. This has enabled them to keep herds of cattle and sheep again, a tribal necessity of which Germany cruelly deprived them.'

The mandated natives are even better off in the south-west than are the natives in the Union of South Africa; the former, for example, have to pay none of the poll tax which often forces the Africans into industry, where their wages never rise. Their tribal institutions have been restored, and the chiefs share in the responsibility of government. The whites who now direct the country are from the Union of South Africa and are, therefore, experienced in dealing with the type of land of the south-west.

No gross cruelties now occur, and, and if any were to occur, they would be reported to the Permanent Mandates Commission. The general result has been a notable increase in prosperity.

In the Cameroons under the French Mandate, there has been a similar approach, carried out in a way that is characteristic of the French. Social services, State services, the construction of roads, education, public health, co-operative agriculture – these all emanate from the top. The French have no faith in ancient systems and traditions; these they destroyed in the Revolution of 1789. They have not, therefore, set about strengthening the status of the tribal chiefs or restoring tribal institutions. As Steer says:

> 'Governors sack chiefs on any pretext. Tradition to them is a thing that clogs.'

With the network of roads which they built, the French introduced efficient communication and transport, enabling them to establish centralization in place of old local administrations. Within these limits, they give the Africans freedom, public service, instruction and medical aid.

Under the Mandate, French authorities have to publish, for all who wish to read, their annual report to the Mandate Commission, and it is due to this, says Steer, that the French Cameroons have an advantage over their neighbour, non-Mandated French Equatorial Africa, whose reports are seldom read outside the Colonial Ministry.

The officials of the mandated areas are spurred by the knowledge of the coming publicity. The consequence is that, whereas the Cameroons grow in strength and population, Equatorial Africa wanes in both. Under eager but careful officials, the freed natives are themselves infused with energy and zeal, and take a growing part in production. Steer writes:

> 'Production is balanced between white and black... Some crops, such as cocoa, which were exclusively European, are now exclusively African... Production has become so popular a pastime that the sources of white labour have dried up.'

This energetic spirit has advanced hand in hand with prosperity, with the result that:

> 'the French Cameroons balance their own budget; they have only in the 1930s borrowed money from France, and their total debt is infinitesimal. Neighbouring Equatorial Africa is one of the territories which promotes most gloom in the French Colonial Ministry. There is always a colossal deficit. Sometimes it amounts to 30 per cent of the total receipts. There is always a crushing debt.'

The British in Tanganyika have worked in the opposite way to the French. Sir Donald Cameron, who knew, as no one else, where, how and to what degree self-government could be developed, gave this succinct account of the British methods:

> 'We built from the bottom, from the common people, upwards on a purely democratic basis, in distinction from other countries, where the tendency has been to invert the pyramid and build from the top.'

The spirit of the Mandates was most acceptable to Cameron. Steer's words on this are so useful that I include them in full:

> 'The Mandate for Tanganyika, as for the Cameroons, destroyed the system of forced labour set up by Germany. The Mandate assured the natives of their primary rights in the land of their ancestors, and security of their interests where they conflict with those of European settlers. The German policy had been the opposite; the Mandate has constructed a peaceful native peasantry where none existed before; it has replaced armed repression with peace, and rough justice with fair and impartial laws.
>
> Above all it has reestablished the natural foundations of native society which had been thoroughly oppressed by the German machine. It found an oppressive and foreign rule in Tanganyika 20 years ago; in that brief time period it has not only restored the native system, but given it responsibilities of which it never dreamed before. The native authorities, suitably democratized, spend their own money, hold their own courts, carry out their own measures of education, hygiene, and all the

> other responsibilities of local government.
>
> There is no need to compare Tanganyika with Kenya or Nyasaland on her northern and south-western frontiers.
>
> Through the breadth of Africa south of the Sahara, you will not find a territory where the native African has such freedom of self-expression as in Tanganyika; or such grand responsibility; or responsibility so faithfully borne. When one lifts the veil of wishful thinking and asks, in a clearer atmosphere, what is the purpose of colonial government, clarity demands no stupid answer such as raw materials for the mother country (seeing that all the colonies of the world produce only 3 per cent of the world's raw materials), or pockets for European investment (seeing that the beneficiaries can only be the few), or strategic power (seeing that if war is permanent, life is not worth living).
>
> No; the light shines too hard today to admit evasion. If we are to remain in tropical Africa we are there for the benefit of the people whom we rule; and their benefit is not only to learn and be healthy, to have peace and to produce; the greatest gift we can offer them is the right to manage their affairs.
>
> That is why Cameron justly said "We have given the people back their soul".'

The Mandates in the three countries have been carried out in three varying ways by the Union of South Africa, the French and the British, but each method has been inspired by the new spirit. It is this that has led to their success and prosperity, for Tanganyika too is prosperous, the richest of the former German Colonies.

The, Mandates, indeed, have worked a miracle of social benefit and civilization in these three countries, especially in Tanganyika. It is impressive, and will become yet more so to the reader of *Chapters 21* and *22*, in which similar miracles on a grander scale will be described.

19

Russia, South Africa, Australia

Russia

The land of European Russia is not complicated. It is the direct extension of *Belt No. 1* of Asia, as described in *Chapter 6*.

Its coasts are too limited and removed from the open ocean to alter its essential character, which, with its situation between Asia and Europe, has directed the history of the Russians. Its physical map is mostly tinted green, indicating an elevation up to 500 feet. It has two irregular areas of yellow, signifying elevation up to 2,000 feet, running north to south; one to the west, the second to the east. The eastern is intersected by a thin strip of light brown, of elevation up to 5,000 feet – the Ural Mountains. These western and eastern areas are joined at latitude 60 degrees by a yellow band. The three areas together form the watersheds of Russia's rivers.

Two considerable rivers, the Dwina and the Petchora, open into the Arctic Ocean, but the largest Russian rivers, unlike those of Siberia, run south. The Volga rises from all three watersheds – west, east and transverse – and runs into the Caspian, its last section in the Caspian Tract being below sea level. The Don originates in the eastern side of the western yellow area, and the Dnieper in its western side, and also from its big tributary, the Pripet, from the northern Carpathian Mountains outside Russia. Both flow into the Black Sea. The Dniester, also entering the Black Sea, forms the south-western boundary of Russia.

Russia is thus an extension of the Siberian Plain, made European by the Ural Mountains. South-east Russia, with the Caspian Tract, is the European extension of the Asiatic Kirghiz Steppes. Through the Steppe country many nomads of Asia passed into Europe and, at a later time, Russians passed into Asia. At the time of the last glacial age, nearly all of European Russia was covered with ice. As the ice receded, Russia emerged in a sodden condition; boggy lands and lakes, the abundant waters of which were drained away by the great rivers. In the drying up of the post-glacial epoch, Russia slowly became suitable for human habitation.

European Russia is now divided into three belts.

Firstly there is the northern belt, with its Arctic tundra cap, a long winter, a brief summer and a saturated, boggy soil. It offers only the most limited opportunities for farming.

The central belt has a more equable climate than the other two. Its soil is capable of receiving and storing water to a considerable depth, and it gets an abundance of water in the spring from the melting of the snows. Its surface then becomes a sea of mud, but during the next five or six months the ground dries, and so cultivation becomes possible. This central belt, which contains the capital and other manufacturing towns, constitutes the farmed, but nevertheless food-deficient, area of Russia.

For these two deficient belts, Russia is compensated by the third or southern belt, the food producing area. Here the season of freedom from snow lasts up to nine months. Here also the extremes of heat and cold are greater than are those of the central belt. So dry is it at times, from the heat and the hot winds which sweep into it from Central Asia, that it is subject to drought. But its soil is rich, and that of the Black Earth Zone, immediately south of the middle belt, is the granary of Russia.

Russian agriculture began in the middle belt, the belt of forests. It was not until the reign of Ivan IV (1533–84) that the grain-producing areas were reached. Sir Bernard Pares, in his most instructive *A History of Russia* (1944), states that the Russian peasants were peaceful, seeking to cultivate land without interference. They would clean a piece of land along the bank of a river, burning down the trees and digging out the stumps. They would erect a colony of a few houses, keeping close to the river for fishing and transport. In the north they met the Finns, then entirely unorganized, with whom they established friendly relations. Pares says:

> 'The Russian peasant was a man of peace, and he did not come to start new conflicts, but to avoid them.'

But they were not left in peace by their rulers in Kiev and Moscow for the simple reason that they were the source of the greater part of the wealth of the country, and also the bulk of the army.

The chief enemies of the Russian people from about 1130 A.D. to the reign of Ivan IV were the nomadic Mongols from Asia, who passed into Russia through the Caspian Tract. To preserve themselves and their peoples from the Mongols, the Russian rulers relied on the strength of the land and its peasants. When Kiev was the capital, the central areas were in the hands of the landed aristocracy, while the distant regions were occupied by pioneer peasants. When Moscow became the capital, land was given to a second class of landowners, who were allocated it on condition of rendering military and other service to the State. Pares writes:

> 'The conditions of tenure stated precisely the number of recruits who had to be provided. The allotment of land corresponded to this number, and was graded carefully according to class, which was practically synonymous with military rank.'

He adds that military service was also demanded from the hereditary landowners:

> 'the patrimonies themselves were put under the same obligation of military service; in fact, *all* land in Russia came to be held only through the title of service to the Tsar.
>
> As a system of agriculture, nothing could be more unsound. The squire was firstly a fighter, and only secondly a farmer. His absences were frequent. His efficiency was rated only by his military service.'

As regards the landowners of the second class, they could be moved at will, which prevented them from acquiring any permanent interest in their peasants or their farms.

The peasants were bound to the soil, for the aggression of the Mongols

> '...convinced the dullest peasants of the necessity of national defence and national sacrifices.'

Nevertheless, some peasants were bold enough to seek freedom from the oppression of their rulers, and moved eastwards and to the south and south-east. Sir John Maynard calls them 'flitters' in his *The Russian Peasant* (1943). They were to become the Cossacks, or border peasantry, who were so powerful a factor in the eventual repulsion and conquest of the Mongols. The Cossacks came to possess, writes Pares,

> '...a wonderful military sense and were masters at taking cover. They practically never parted with their horses and were trained riders from childhood. Their scouting tactics were those of the Russian army of today. Tall trees were used as observation posts; different points at some distance from each other were garrisoned, and between them relays of individual Cossacks patrolled, never dismounting.'

The Cossacks, because of these qualities, enjoyed a freedom not given to the more central peasants.

Even when the Mongols were finally subdued by Ivan IV, the heavy burdens placed on the Russian peasants were not lifted. The danger to the Russian kingdom had shifted from the south and east to the west, as Europe advanced both in civilization and military strength. The virtual serfdom of the peasants, in degree severe according to their relative proximity to Moscow, was continued, and in 1649 a code of laws was issued which finally confirmed the establishment of serfdom, which from that time on became a state institution.

Not even Peter the Great (1682–1721), who in physical strength, will, genius and energy was perhaps the greatest of all European rulers, could relax the oppression of the peasants. It was he, above all, who realized that Russia would be lost in the struggle against European aggression unless her backward people were Europeanized. So, although Peter was far closer to the Russian peasant than any Tsar before him or since, his vast expenses forced him to increase the weight of the peasants' burdens.

'Flitters' to the freer south and south-east increased greatly in numbers, and some passed over the Urals into the Siberian Plain, which was familiar in character.

As time passed, the freeing of the peasants from serfdom in order to promote the advancement of the largest part of the population became more and more urgent an issue, and eventually an *Act of Emancipation* was passed in 1861. It was in many ways ineffective.The next big development came with the Soviet Revolution and the collective farms of Lenin and Stalin. Under the collective farms, the medieval tempo of farming was transformed into modern farming of an advanced kind. Sir John Maynard, in his sixteenth chapter, gives a full account of these collective farms, with their advantages and difficulties, their tractors and machinery, their economics, their lack of sufficient manure, and the nature of work on the large farms compared to the private farms.

*

Two forms of erosion have occurred in European Russia.

The first occurred through the destruction of the forests on the slopes and watersheds of the higher land, when the landowners were anxious to get the most wealth from the soil with minimal return. When the forest had been destroyed on the slopes and watersheds, sheet and gully erosion would occur with the onset of heavy rain or melting snow. The rivers would, in the wet periods, then flood and carry top soil to the sea.

Jacks' description in *The Rape of the Earth* of this type of erosion is one with which we are already familiar. The peasants of Russia were serfs; all the profits of farming went to enrich the landed aristocracy. We are told:

> 'The landowners' aim was to get the maximum out of the soil in the shortest time, with the least expenditure of labour and improvements. Sheet erosion was extreme, though generally unnoticed and not associated with the new gullies that continued to break up the land.'

Fortunately for Russia, the primitive farming of the serfs was slow in its expansion. Though marshes and lakes dried up and the streams of great rivers flooded and dwindled, the overall tempo was very much less than that of the United States under the era of the machine. Consequently, the loss of the Russian soil was far less than the disastrous magnitude of that of the American soil.

When the serfs of Russia were emancipated in 1861 and became the owners of much of its land, there was no halt in this form of erosion. The peasants were given the poorest, most eroded land, and they therefore set to work to extend the cultivation of the slopes. Their method was to plough long strips up and down the slopes, to the tops of the watersheds; the hollows between the ridges became first watercourses and then gullies, and so erosion was increased.

Under the Soviets, the destruction of forests continued, because of the need for timber to trade for the machinery which the new manufacturing towns of the USSR required.

So much for the first form of erosion, namely, sheet and gully erosion. The second form is wind erosion.

This affected the third or southern belt of Russia, both the rich black soil zone and the land of the Steppes. Here is what Major Law, a commercial attache in Saint Petersburg in 1892, said about this region, once protected by belts of forests:

> '...those forests do not now exist, and the black soil country is often scourged by devastating blasts from the Steppes, and not infrequently baked by prolonged droughts. Wind erosion and floods work their havoc and smite the soil with perpetual barrenness.'

This erosion at length aroused alarm in the Russian aristocracy and, in the 1890s many shelter belts of trees were planted to break the force of the hot winds upon the top-soil,

> '...on such a spectacular scale and with such excellent results that a special government commission was appointed to study reforestation.' (Whyte, *The Rape of the Earth*).

The sudden entry of the USSR into the modern era, with its reliance upon machines, made the tractor in particular a symbol of modernization. It became the visible image of the belief in the machine, and wherever the tractor went it was heralded with much fanfare.

Nevertheless, it had its intrinsic dangers. Jacks and Whyte, in *Technical Communication No. 36 of the Imperial Bureau of Soil Science*, 1938, discussed this danger and quoted the Russian Professor Kornev as saying, with regard to both forms of erosion:

> 'At the present day there are huge areas in the USSR where, owing to the excessive breaking up of the topography, whole territories, formerly under profitable agriculture, are now occupied by immense ravines and infertile wastes.
>
> The tractor-plough is the enemy of the grassland in dry areas, but is indispensable to the propagandist of Russian agriculture. Though forewarned by the experience of other countries, it is difficult to ascertain whether the authorities are aware of the danger of mechanization.'

By 1938, the Soviet authorities had put tractors on the land on a vast scale. Whereas, according to *Appendix III* of the Trade Unions of the USSR (quoted in the Fabian essay *Our Soviet Ally* (1943)), in 1913 no tractors were in use in all of Russia, while in 1938 there were, upon the scientific, mechanized, collective farms, no less than 483,000 tractors in use. The potential threat of erosion resulting from such numbers is enormous, as the consideration of our other three examples of modern erosion – South Africa, Australia and the United States – will show.

The war and its demands have distracted attention from the threat in the USSR, and even made it one that has had to be concealed. No account of the degree of new erosion can, therefore, be given.

South Africa

South Africa is dealt with by Whyte in *The Rape of the Earth*, under the heading of *The Transformation of South Africa into Semi-desert in the Twentieth Century.*

The agricultural wealth of South Africa is chiefly pastoral. The natural veld and the Karroo provide animal fodder, though in favoured localities special grasses and foreign crops are grown. But this natural vegetative cover has deteriorated, and

> '...erosion has already transformed parts of the richest pastoral areas in the country into semi-desert. Considering that the luxuriance and excessive wetness of the veld in the Orange Free State were previously an obstacle to pastoral farming, the rapid appearance of the disastrous consequences of erosion is very remarkable. It occurs in all parts of the Union, either as an actual or probable menace, and is predominantly a pastoral problem.'

He continues with the following observation:

> 'The great uncertainties of the South African climate, and the suddenness with which the country was opened up after the discovery of gold, have contributed largely to the rapid acceleration of erosion.
>
> Towards the end of the nineteenth century it was realized that serious overstocking was taking place, but public attention was not focused on the danger until the Drought Investigation Committee issued its final report in 1923. Until then the opinion had been gaining ground that the climate was becoming drier and the rains, when they happened, more torrential.
>
> The report pointed out that there was no proof of definite and recent climatic change, but that erosion would account for the drying up of

> rivers and waterholes, the falling watertable and the increasingly disastrous effects of droughts and heavy rains.
>
> The Commission concluded that this erosion was caused chiefly by deterioration of the vegetative cover brought about by incorrect veld management, and that all efforts to improve the latter would have a beneficial effect on the former.'

South Africa was taken from the Dutch by the British in 1812.

Australia

In the speed with which the fertility of its soil is being lost, Australia is believed to surpass even the United States. This is the opinion of Mr E. Clayton of the Department of Agriculture, New South Wales, who was sent to study erosion and anti-erosion measures in the United States for the better defence of his own country. In *Investigations Overseas* (1937), he writes:

> 'There is no doubt that we Australians are in a process of transforming the semi-arid areas into desert at a more rapid rate than in the USA.'

Australia is, in short, being threatened with becoming to the British Empire what Libya became to the Romans. The loss of soil from rapid deforestation, burning of vegetative cover and overgrazing is severe. *Bulletin 13 of the Commonwealth Bureau of Forestry* describes the catchment area of Australia's greatest river, the Murray:

> 'Approximately two-thirds of the area covered by Alpine woody shrub has been completely cleared by the action of fire.
>
> The organic layer, with no cover to protect it and no live roots to hold it, dries up and is blown away; the loose sandy soil is in its turn blown

> away, leaving the final product – bare granite rocks and stones with no vegetative cover.'

The once constant river has become inconstant, and its water is intermittently turbid in place of its original clarity.

> 'In 30 years the land about it has become desert, according to the testimony of men who have nurtured cattle there all their lives.'

The eagerness for wealth in a country that does not nestle to the heart as does the homeland destroys the permanence of its gifts to humanity. There is no real ecological link between most of the white farmers and nature, nor a tradition which expresses it. They do not feel that when they burn or tear at the soil they are tearing at their own homes, at something which is an eternal associate of their own, and their ancestors' and their descendants' lives.

It is land, and land of course can give wealth, but it is not *motherland*, and, until that term and what it implies in the fullest sense becomes bred in the bone, it will not be real, and the land will not be properly treated.

It is the heart, faith, and sentiment that ultimately prompt action. So in the drier parts of Australia there are thousands of acres of lightly stocked pastoral country, which are suffering from erosion and where men, like fifth columnists, have helped the central desert to advance. Not only is the fertility of these undulating lands being depleted at an alarming rate, but the wetter, riverine districts near the sea are also in many parts gravely affected by erosion, due to the same eager speed at which land is being cleared, as has already done so much harm to the riverine area of the Murray River.

Finally, there are the rabbits, introduced from Britain. In *Pamphlet No. 64 of the Commonwealth of Australia* there is a vivid picture of these pests. In a way, these four-legged immigrants from Britain behave like the two-legged variety. When the going is good and there is abundant pasture, the rabbits do not act as if they were part of a country in balanced equilibrium, but as pitiless ravagers of its soil fertility. They revel in the rich harvest and they multiply beyond all calculation, as if the future is but a repetition of the past. In huge numbers, they eat up the pasture and force hungry sheep to devour saltbush. They overreach their good fortune, and then come the hard times of a drier season. In their hunger they eventually eat all and any food within reach – and even beyond the ordinary reach of their kind. They eat the surface plants. They climb. They burrow into the earth and get at the roots of the hardy acacia scrub. They take all that is above and below the soil and give it no chance of regeneration. Then comes a drought, and the rabbits die in heaps under the eaves of the settlers' houses or wherever manmade shade can shelter them from the pitiless sun. So the rabbits, too, fail to fit into the delicate balance which nature has long established in Australia.

The British first settled in Australia a century and a half ago, and annexed New Zealand (the islands of the 'emaciated skeleton' threat mentioned at the end of *Chapter 17*) in 1840.

20

The United States of America

OF ALL THE COUNTRIES in the world, that which most typifies modern progress – that which at its foundation proclaimed the liberty of every man to pursue wealth within the limit of the law, and has permitted the greatest freedom of thought and action on the part of its inhabitants of all ranks, is the United States of America.

The USA has encouraged its citizens to develop themselves freely, in the belief that the knowledge and wealth which accrue as a result are of ultimate benefit to the community and to humanity as a whole. It has become the nation with the most advanced scientific equipment and practical techniques, and has rushed forward into the new era with such speed and eagerness that the old has been forgotten.

But upon the USA, Nature has now written most broadly and definitely her judgment – through erosion of the soil. And yet, with its usual heroic practicality and vigour, America is now surpassing other nations facing similar problems in the thoroughness of its measures to oppose this life-destroying menace.

In April 1928, the Agriculture Department of the USA published *Circular No. 33*, entitled *Soil Erosion: A National Menace*. It is divided into two parts; a general consideration of the loss of soil due to erosion, and a discussion of the erosion of grazing lands. It consists of 35 pages and contains 35 photographs, and is a graphic revelation of the greatest rebellion of our time – the rebellion of the Earth itself.

The figures of destruction given are colossal, and when one looks at the photographs and sees these figures embodied in pictorial form, the effect is so impressive that one can understand how this short pamphlet assisted, as it did, in uniting the voices of so many.

One of the authors discusses soil wastage:

> 'The amount of plant food in this minimum estimate of soil wastage by erosion (1,500 million tons of solid matter annually) amounts to 126,000 million pounds, on the basis of the average composition of the soils of the country, as computed from chemical analyses of 389 samples of surface soil collected by the Bureau of Soils.
>
> This is more than 21 times the annual net loss caused by the harvesting of crops. The amount of phosphoric acid, nitrogen and potash alone in this annually removed soil material equals 54,000 million pounds.
>
> Not all of this wasted plant food is immediately available, of course; but it comes principally from the soil layer, the main feeding reservoir of plants, and for this and other reasons it is justifiable to consider the bulk of it as essentially representing lost plant food, without any quibbling about part of it having potential value only.'

It is impossible to comprehend fully figures that are so colossal. There are 120,000 million pounds of plant food lost by 120 million people. The permanent loss of plant food, therefore, is at the rate of 1,000 pounds per person in the USA. As each person eats about 1,000 pounds of food a year in either plant form or animal form derived from plant substances, one can get some sort of equivalent conception as to what this means.

Nor is this all. The destruction is proceeding at an

ever-increasing rate, as we shall see in the following quote. Clearly such a system of treatment of the land cannot continue indefinitely. One recalls the prophetic words of Professor Shaler of Harvard, uttered 30 years before the publication of this pamphlet:

> 'If mankind cannot design and enforce ways of dealing with the earth which will preserve the sources of life, we must look forward to a time – remote it may be, but clearly discernible – when our kind, having wasted its great inheritance, will fade from the earth because of the ruin it has accomplished.'

There is also something paradoxical about these incomprehensibly large figures, attached to something so general to humanity as the soil, that connects them to the money system of the same period. Farmers create what men eat, but in doing so create life-strangling erosion and deserts. In the money system, financiers create money, but in doing so they also create life-strangling debts and financial deserts. The vast accumulation of negative money, as national and municipal debts, runs into figures comparable to those that describe the deplorable state of the soil.

The two systems seem to have a kinship. Yet the urban population still view the scecurity of both their food and their money with a scarcely shaken trust. Again, the split mind gives evidence of itself.

There are further figures. The loss of phosphorus, potash and nitrogen alone, without reckoning other soil foods, is estimated at 2,000 million dollars a year, which is the better portion of the British national revenue before the war of 1914-18. Mr Bennett, writing in 1928, which he denotes as a time of lack of fundamental data concerning 'what is going on at an ever-increasing rate', states:

> 'That some 15,000,000 acres or more of formerly tilled land has been utterly destroyed by erosion in this country is but an insignificant part of the story, for it is the less violent form of erosional wastage – sheet erosion – that is doing the bulk of the damage to the land.
>
> Land depreciation by this slow process of the removal of the top soil is of almost incalculable extent and seriousness, and since the denudation does not cease when the subsoil is reached, there must be in the near future – unless methods of land usage are very radically changed – an enormous increase in the abandonment of farms.'

Nine years later, E. Clayton of the Agricultural Department of New South Wales, after his study of erosion in the USA, reported back with these momentous figures: 50 million acres of cultivated land destroyed, 50 million acres seriously eroded and about to be abandoned, 100 million acres with loss of much of the topsoil – out of 987 million acres, the total agricultural land in the States.

Finally the United States Department of Soil Conservation published a map that uses shading to show areas of slight and severe wind erosion, and slight and severe sheet erosion due to widespread movement of thin sheets of water. The unshaded (non-eroded) areas contrast with a widespread prevalence of shade. The above figures of Clayton are certainly not belied by the map.

The figures of loss in the USA are, indeed, incredible. The mind can barely grasp that a country of such vast national wealth and fertility can be thrust into such danger in the course of a single century. Here is a paraphrase of Mr Bennett's own summary:

> 'To visualize the full enormity of land impairment and devastation brought about by this ruthless agent is beyond the possibility of

> the mind. Any American with any imagination knows that the people of the United States would willingly spend twenty billion dollars to redress the wrong, had it been due to a foreign foe.'

But because it is an inherent fault in American thought, and because the sun, the wind and the rain – the natural conditions of earthly life – are concerned, the people scarcely heed it. *Circular 33* concludes with the words:

> 'A little is being done here and there to check the loss – an infinitesimal part of what should be done.'

Then comes the visual evidence; photographs from various parts of the USA. In the first photograph one sees a sloping cotton field showing shallow channels caused by rain between the rows of plants. When heavy rains came and the field had only a light cover of young plants, more water ran along these channels between the plants, rather than where the soil was held by the roots of the plants.

This is sheet erosion; some water sinks into the soil, but much of it runs away without sinking in. The runnels collect together and form a gully. The next photograph shows the result of these gullies, which have collected together to form one huge channel in the fine sandy loam. The channel is a chasm 100 feet deep, with steep sides and fringed by forest trees. Where it now is, a school-house stood just 40 years ago.

There follow further pictures of erosion due to rain and melting snow. There is one of a wide, laterally extending gorge in the Mississippi Valley; one of cornfields covered with a blanket of coarse sand deposited by the action of heavy rain; arable land in Kansas so cut up by gullies that it could no longer be ploughed and was given over to pasture; bald patches on the rich, black soil of Iowa, washed away by sheet erosion and showing the clay beneath; smooth fields in Texas split by gullies just as a flat glacier is split by crevasses; rolling hill country in Northern California, once forested but now complete desert, with no topsoil and what is left of subsoil slashed by gullies; a spacious hilly area in California left desolate of growth by the combined action of fire and water; in Virginia erosion following a few drops on slopes which should never have been cleared; in Colorado, farm buildings caught and undermined by a wide stream, the natural obstructions to the free flow of which the owners of the buildings had removed; wagon tracks starting eroding streams, which will eventually lead to the loss of a valley full of rich soil; driftwood and other debris from hills made barren by fire and deposited by flood upon a young orchard; wind erosion due to trampling of an excess of cattle in dry New Mexico; and vegetation destroyed by smelter fumes over thousands of acres in Arizona.

The last picture of all, though, is one that cheers the heart after so many images of destruction. It is one of 'abundant and excellent feed and a maximum of watershed protection' in Montana. It shows a well-watered, hilly country bearing tall fir trees, bordering spaces of rich grass, and in the foreground, a flock of feeding sheep. The land and its cover are good and humanity has joined the local life-cycle in a state of harmony.

Nevertheless, Montana also takes its share of the famous grazing grounds of the north-west, where it is said that 58 million acres are now only able to feed *one fifth* of the number of animals they were once able to support.

It is in Montana's neighbour, the Pacific North-West, that something very positive has arisen, as a result of new values in keeping with the awakening of some Americans to the primal value of the soil, protected by the continuous ownership of those who will care for it on small-sized farms. Once again, the land is to be a homeland, and not just a commercial possession.

In the Pacific north-west flows the great Columbia River, across which have been placed two dams; the Bonneville and the Grand Coulee. These two huge dams will produce more electric power than all the 14,082,282,000 kilowatt-hours turned out by the 260 electric plants in the State of New York.

What to do with this enormous amount of energy? There enters into this question something new to industrialists. They wanted a vast factory community near the Bonneville Dam itself – but fortunately the USA at that time had a leader who possessed a vision that extended from the present into the future. In September 1937, President Roosevelt delivered a dedication speech beneath the dark crags of the Columbia Gorge, the river of which now promised great change for the American people.

The North-West, consisting of the States of Washington, Oregon, Idaho and the section of Montana west of the crest of the Rockies, offers an opportunity, said the President,

> '...to avoid some of the mistakes and wasteful exploitation of resources that have caused such serious problems in other parts of the country.'

The North-West was not to be a land of new 'Pittsburgs'. The President continued:

> 'It is because I am thinking of the nation and the region 50 years from now that I venture the further prophecy that as time passes, we will do everything to encourage the building up of smaller communities in the United States.
>
> Today many people are beginning to realize that there is an inherent weakness in cities which become too large, and inherent strength in a wider geographical distribution of population.'

The Grand Coulee, now nearing completion, in addition to providing power will irrigate 1,200,000 acres. This land is to be given to families and small cultivators, and will be protected from combines, corporations and other large-scale operations, whose ownership of it is forbidden.

Land held in defiance of this limitation will get no water from government canals. Families who have been driven by erosion from the western grazing lands and who have migrated from the now famous Desert Bowl will find land here; some are doing so now. The amount of land allowed to be held is limited to 80 acres for a family and 40 acres for a single person.

The purpose of the Grand Coulee is to take care of as many families as possible. The partnership of family and soil is to be revived, but there is to be something more on this irrigated land; there are to be small industries served by electric power. People who work at these industries will also be able to have kitchen gardens, and so will create what Stuart Chase proposed several years ago for the North-West: small farms which will act as 'anchors to windward', if at any time industry fails.

Everyone in this area will have the opportunity to learn – or relearn – soil sense. The land as food producer will be the basis of society and will be its associate. Many small industries dotted about the North-West will serve the countryside, as village crafts once did. Other industries will develop the raw materials from the local farms for industrial purposes. Industry will be truly distributed; it will cater to the comfort and happiness of the people on the land as its primary object, and act as the means of external trade only as its secondary object. The dicta, indeed, of the North-West, will be those expressed by Napoleon at Saint Helena:

> 'Agriculture is the soul, the foundation of the Kingdom. Industry ministers to the comfort and happiness of the population; foreign trade is the superabundance, and allows the exchange of the surplus of agriculture and industry.
>
> Foreign trade, which in its results is infinitely inferior to agriculture, was an object of secondary importance to my mind. Foreign trade ought to be the servant of agriculture and home industry; these last ought never to be subordinated to foreign trade.'

The cost of the electric power of the Bonneville Dam is governed by 'postage-stamp rates' all along its transmission line of 275 miles; the industry farthest from the dam pays for its power at the same rate as the industry that is nearest. This mandate was not directed against industries and factories as such, but against industries and factories located in places like Pittsburg, Chicago and Detroit.

The fear of the great metropolitan city is so ingrained in the thought of the people, writes Richard Neuberger in *Free America* (August 1940), that during the struggle over the Bonneville power rates, the words of President Jefferson (1743–1820) appeared in many local papers:

> 'I view great cities as pestilential to the health, the morals and the liberty of mankind.'

That saying was directed against the financial and industrial magnates and ambitious politicians and demagogues who arise in cities, and are made possible only by cities. The logical end of urban civilization – and its most complete form – is *totalitarianism*, which is 'pestilential to the liberties of mankind'.

In this scheme in the Pacific North-West, people will have the opportunity to combine manufacturing with a partnership with the soil. The soil, once again and in the future, will be their associate and instructor. There is a grandeur about the scheme which quite rightly belongs to a great country that can still so resolutely seek to restore its epic character.

There is another illustration of the redemptive spirit in the USA which stirs hope and admiration no less than the story of the Grand Coulee. It is the story of the almost complete re-education of the people in the wisdom of the local soil.

It was inspired by the great catastrophes of the Dust Bowl and the floods in the basin of the Mississippi River, and eventually united all classes of the inhabitants there. It is described in a pamphlet issued by the U.S. Department of Agriculture in October 1940.

Pupils and teachers used the local land as their textbook. They learned to recognize the symptoms of misuse, to discover the causes, and to work out principles of good use. They studied wind erosion and the Dust Bowl. The results of flooding and the subsequent loss of valuable cotton land led to the study of the watershed and its interrelationships, and of the delivery of water from the forested areas to the irrigated land below.

Because of the war I have not as yet been able to procure the pamphlet itself, but I have read a review in *Indian Farming* (January 1942), which quotes verbatim from the pamphlet. It is so important and encouraging that I am reproducing the quotation in full:

> 'Basic concepts and bodies of subject matter were needed – an understanding of the water cycle, the behaviour of the soil and water, and the growth of vegetation. These were observed and understood and related to the daily life of human beings.
>
> Children gained some understanding of the water cycle in the simple story of the raindrop.

Grass as a necessary food for livestock was known to even the smallest child in the south-west. How grass grew, how it reproduced, how over-grazing and trampling destroyed it, led quite logically to such statements as: "The cowboys should not let the cattle eat in one place too long." Sustained use of timber from forested land led to the acceptance of the necessity for large trees, middle-sized trees, and little trees. Human use, human needs, human plans and solutions, were the core of each study.

Children have a way of talking about matters that really interest them. Visits by pupils to demonstration areas have led to visits by parents. Parents have written letters to schools expressing their interest and pleasure upon learning that the children are studying land use. In sections where this type of education was going on, the technical men reported an added interest in the districts and a great facility in obtaining agreements (presumably for the better use of land and water by farmers and local authorities).

The educational superintendents, supervisors, and departments of education lent every facility, gave advice, and even took over where possible. The technical staff of the Soil Conservation Service conducted tours, learned to adapt their language to children's understanding, and frequently wrote expositions in lucid, simple language.

Teachers, recognizing that soil conservation was of great interest to their community and that it helped in the vitalizing and socializing of the whole school programme, threw themselves into the new projects with originality and eagerness.

Our brief experiment has shown that land planning and use has an immediate interest for

> every school, and that teachers, pupils, parents, and State officials are eager to have a part in it. It is one of the great problems before us today. It has to do with subsistence, with food, clothing, shelter, taxes, and with many other problems which are a daily part of the home, community and the nation.'

Everyone concerned, it must be noted, becomes interested. It is a call to all from their very origin itself, and each man, woman and child eagerly respond to it. It is nothing less than the construction of children's minds – and the reconstruction of their elders' minds – in keeping with the wisdom of the soil.

Mr Bennett sums up his survey of the soil of the United States with these fateful words:

> 'After 4,000 years of building dykes and digging great systems of canals, the Yellow River broke its banks and killed a million people during a single great flood. During that flood the Yellow River, known in China as "the scourge of the sons of Han", changed its course to enter the sea 400 miles from its former mouth.
>
> No one, of course, wants anything remotely like this to take place in this country, but coming events cast their shadows.
>
> That the greatest flood of which we have reliable records came down the Mississippi in 1927 was a prophetic event.
>
> G. Martin's statement about erosion as an enemy to agriculture – "It is very unlikely that any other industry could suffer such losses and survive" – is also prophetic.
>
> That bare land at the Missouri Agricultural Experiment Station was found to be wasting 137 times faster than land covered with blue grass on a slope of less than 4 per cent gradient is prophetic.

> That many millions of acres of land lie bare and desolate and exposed to the ravages of fire and erosion, with but pitifully little done towards reforestation, is prophetic.
>
> That estimates show that the rate of plant-food wastage by erosion is 21 times faster than the rate at which crops are harvested, is prophetic.
>
> These shadows are portents of evil conditions that will be acutely felt by posterity. Should we not proceed immediately to help the present generation of farmers and to conserve the heritage of the land?
>
> The writer, after 24 years spent studying the soils of the United States, is of the opinion that soil erosion is the biggest problem confronting the farmers of the nation over a tremendous part of its agricultural lands. It seems scarcely necessary to state the perfectly obvious fact that a very large part of this impoverishment and wastage has taken place since the clearing of forests, the breaking of the prairie sod, and the over-grazing of pasture lands. A little is being done here and there to check the loss – an infinitesimal part of what should be done.'

These words did not fall on deaf ears. The President and Congress were deeply stirred, and five years after the publication of *Circular No. 33*, the Tennessee Valley Authority took control of the valley of the Tennessee River and its tributaries, an area belonging to seven different States, and of no less size than England and Scotland combined.

This was the first answer of the Government of the United States to the question of the circular: "Shall we not proceed immediately to help the present generation of farmers and to conserve the heritage of the land?" It was the first real response to the fact that 'little that is being done here and there to check the loss – an infinitesimal part of what should be done'.

To help in understanding this great project, it is advisable to recall how in Tanganyika – in order to avoid the erosion caused by the wholesale destruction of forests harbouring the tsetse fly – geologists, plant ecologists and water surveyors were called together to adapt local farming to the character of the water supply as a whole. Each river with its catchment area was made into a locally governed entity, and 26 such entities were combined into one Union.

What the tsetse fly forced upon the discerning mind of Sir Donald Cameron, the devastation and poverty of the Tennessee River area also brought to the attention of President Roosevelt and a strong following among the members of Congress. The story has been told with a comprehensiveness worthy of the theme by the Chairman of the TVA, David Lilienthal, in his book *The Tennessee Valley Authority* (1944).

The story begins with the natural unity of the Tennessee Valley area, with its forested catchment areas of mountains and valleys, and the varied animal and vegetable life they support. The forest protected the soil against heavy rainfall by allowing the rainwater to filter through the soil, and return via the Tennessee River to the Mississippi and then the sea.

Then came the white settlers. They surveyed the land and found it suitable for two good money crops: cotton and tobacco. There were also many trees on the mountainous ridges and slopes which provided valuable timber; there were minerals worth smelting; and there were swift, clear rivers, which, if harnessed by dams, would yield electric power.

These primeval mountains and valleys were full of promise. The settlers set to work, each individual or group following their own plan, and soon the primal unity of the valley was destroyed.

At first the land seemed to be indeed a land of promise, but of course it was abused, and finally rebelled. The time came when the hill farmers found their land scored with gullies, and the farmers on the plains had their fields coated with silt from floods. The owners of the timberlands, neglecting reforestation, saw their stock depleted and barrenness take its place. Hardwood fuel was no longer enough to serve the ore smelter furnaces, and the fumes of their ovens killed the thin vegetation which attempted to cover the deforested land. Finally, the owners of dams found their pipes blocked by the silt carried by floodwater, and electricity no longer flowed so abundantly from the dynamos which the piped water drove. The river itself was thick with silt, local navigation upon it was impaired and, in flood, farmland was washed away.

Each of these problems compounded the others, so that before the Tennessee Valley Authority took over the whole valley, the inhabitants were the most poverty-stricken and backward of any people in the USA. They were in the front rank of those whose destructive activities were described by Mr Bennett in the results of his 24 year study of the soils of America. The outlook was as ominous as it could have been.

The sole hope was to alter the very principles and methods of the use of the valley as a whole, and to reintroduce those of the unity of nature, which had been ignored and fragmented. This was the work which Congress delegated to the Tennessee Valley Authority on 18 May, 1933.

In the brief space of ten years, the TVA has erected sixteen dams, some of them amongst the biggest in the world, and taken over and modified the five existing dams, and made them into one system of regulation of the rivers under their central control. The TVA now controls the entire river system. In 1942, when torrents of floodwater came raging down a large part of the catchment area of the mountains, they were conducted safely into controlled channels, protecting both the Tennessee Valley and its 4,500,000 inhabitants.

The TVA has planted over a million trees that are compatible with local conditions. It has introduced contour ploughing, terraced cultivation, farmers' woodland, and a balanced economy of legumes, clover, pigs, poultry and cattle – and introduced the concept of crop rotation – on 20,000 demonstration farms, of the 225,000 farms (with 1,350,000 people living on them, in family farms averaging 75 acres in size) in the area.

The TVA also manufactures phosphates, proclaimed by its experts as being the most needed artificial manure, and Mr Lilienthal calls it 'the almost magic phosphate' on account of its results. They have, with their mighty dams, created cheap electric power which gives each person 2,400 kilowatt hours (compared to the average for the USA of 1,530 kilowatt hours). This has increased industrial capacity, and as a consequence in 1943 the USA was able to build its huge fleet of bombers for use in Europe and the South Pacific.

The TVA has also made a stretch of 464 miles of the river navigable, with a depth of six feet, and they will soon have a stretch of 650 miles with a depth of nine feet. Finally, the number of fish in the river and its reserves has increased fifteen times.

To effect these changes, the TVA has had to adopt new principles. They had to consider the interrelationships and independence of the different aspects of nature, instead of seeing nature as a battle in which each element of the life cycle, be it animal, vegetable or mineral, is set against the others in a struggle for survival and priority.

Mr Lilienthal's book illustrates and discusses, in a variety of aspects, the change which was necessary, not only in Congress itself, but in every inhabitant of the Tennessee Valley. Here is some of what he said:

> 'Congress, in creating the TVA, broke with the past. No single agency had ever been assigned the task of developing a river in order to release the total benefit of its waters for the people.
>
> The *TVA Act* was nothing inadvertent or impromptu. It was the deliberate and well-considered creation of a new national policy. For the first time in the history of the nation, the resources of a river were to be "envisioned in their entirety". They were to be developed in that unity with which nature herself regards her resources – the waters, the land, the forests together, a seamless web – just as Maitland saw "the unity of all history", of which one strand cannot be touched without affecting every other strand for good or ill.
>
> Under this new policy, the issue of creating wealth for the people using the resources of this valley was to be faced as a single problem. To integrate the many parts of that problem into a unified whole was to be the responsibility of one agency. The Tennessee Valley's resources were not to be dissected into separate bits that would fit into the jurisdictional pigeon-holes into which the various parts of government had by custom become divided.
>
> It was not accepted that at the hour of creation, natural resources had been divided and classified so as to conform to the organization chart of the federal government. The particular and limited concerns of private individuals or agencies in the development of this or that

> resource were disregarded and rejected in favour of the principle of unity. What God had made one, man was to develop as one.'

The TVA controls and bears responsibility for the dams, their electric power, the giving of advice to farmers, the adaptation of industry to the new system, and general supervision and planning.

In all other respects, the greatest possible degree of freedom and responsibility has been given to the people of the valley, through decentralization. 'A man wants to feel that he is important' is the maxim that directs this.

> 'The very essence of the TVA's method in the undertaking was at every hand to use directly, and to encourage and stimulate, the broadest possible coalition of all forces.
>
> Private funds and private efforts, on farms and in factories; state funds and state activities; local communities, clubs, schools, associations, co-operatives – all have had major roles. Moreover, scores of federal agencies have co-operated.' – here a list of 20 is given – 'The list, if complete, would include most national agencies.'

The farmers themselves decide which farms shall be demonstration farms. The distribution of electric power is directed by the farmers, the industries, the municipalities, and the States. The experts live amongst the people and are one with them. Labour is sourced primarily from the people of the valley; others, chosen by merit, are directed to expert work. No inducements are offered to industries located in other regions to move to the Tennessee Valley.

Responsibility is distributed. The TVA management is responsible to Congress, yet it is a separate authority, and its separation is underlined in that neither its management nor its staff are permitted any share in politics except that of voting. The same separation exists between the management and the staff; its members are encouraged to act and take responsibilities, and not worry about mistakes.

A similar relationship exists between TVA and local bodies and associations, who are given – and readily accept – action and responsibility for their localities. Mr Lilienthal himself describes it as "democracy on the march".

The results have awakened the keenest interest, not only in the United States itself, but in other countries of the world. An impoverished and fear-stricken people in just ten years have become prosperous, confident, well fed, and well clothed. They are happier and better citizens.

But one principle of service to the soil is missing from Mr Lilienthal's book – the rule of the return of what is taken from the soil. It is obeying this rule that has, for example, allowed the Chinese to maintain the health and productivity of their soil for so many centuries.

There is one oblique reference to it in the statement that if cotton-seed oil mills made money, Tennessee cattle could be fed with the cotton-meal cakes they now export for sale. If they did this,

> '...as much as 80 per cent of the fertilizing value of the meal would be returned to the soil rather than continuously drained by export.'

Apart from disregarding the rule of return, the TVA is a wonderful rediscovery of almost forgotten laws. How great this rediscovery may yet become we shall now see in the story of a kingdom in Europe – one which was, one might say, the Tennessee Valley, and much more.

21

A Kingdom of Agricultural Art

IT IS REFRESHING – and an essential restoration of the mind – to turn from the dismal state of farming under modern civilization and to consider how a kingdom in western Europe possessed an agriculture achieved the status of art. We have to go back a thousand years to find it. As well as this long interval of time, there are other distances between this great agricultural society and our own European culture.

The race which created this society is no longer to be found in Europe. Moreover, the religious faith which it held to is now only found in Europe in a few mountainous areas. In its racial and religious characteristics, then, this society was strange to Europe, yet in spite of this strangeness – or perhaps because of it – it attained a degree and fullness of civilization not reached by any other European people.

This kingdom was that of the Arabs in Spain, which began with their invasion under Tarick in 711 A.D. and came to an end with the fall of Granada in 1472.

The story of its achievements has been related by S. Scott in the three volumes of his *History of the Moorish Empire in Spain* (1904). It contains a wealth of detail collected during a period of more than 20 years.

Particularly notable is the account, in the thirtieth chapter, of the agriculture on which this flourishing empire was based, and with which it supported a population believed to exceed that of the combined populations of England, France, Germany and Italy of that time. The figures given are 15 to 20 million for the four countries, compared to 30 million in Arabic Spain.

The nature of their agriculture, existing in what were the Dark Ages for the rest of Europe, will be so strange to many readers that it may seem beyond credibility. This is because in English education, the influence of Arabic culture is entirely left out. We are taught a lot about the Greeks and Romans as intellectual leaders of Europe, but nothing about the Arabs.

Scott, though, is well aware of the special situation with which we are now concerned. He anticipates the incredulity with which his account would be met:

> 'In all the vast domain of historical inquiry, there is probably no subject which has been treated with such studied neglect, and with such flagrant injustice, as the civilization of the Arabs in the Spanish Peninsula. Its story has been written in the majority of instances by the implacable enemies of those who founded and promoted it. Theological hatred has lent its potent aid to the prejudice of race, and the envy arising from conscious inferiority, to deny or belittle its achievements.'

A single example will illustrate just how bitter this theological hatred could be. Eulogius, a learned Spanish priest, discovered through his studies – or invented – the fact that Mohammed announced to his followers that three days after his death, he would be raised by the angels to heaven;

> '...instead of this, dogs devoured his rotting corpse.'

This example is taken from the well-known book *Islamic Culture* (1905) by S. Khuda Baksh. When one thinks of the reverence with which Mohammed spoke of Jesus in the Koran and the same reverence which he transmitted to his followers, one can see on which side the bitter religious hatred lay. Consequently, in view of this neglect and prejudice, Scott, under the heading of *Authorities consulted in the Preparation of this Work*, lists no less than 703, covering fourteen different languages. Many of these books are concerned with Arabic culture and history as a whole.

In the sphere of science and thought for example, William Lecky, in his *History of Rationalism* (1865), paid this tribute to the Arabs:

> 'Not till the education of Europe passed from the monasteries to the universities, not till Mohammedan science broke the sceptre of the Church, did the intellectual revival of Europe begin.'

John William Draper, in *A History of the Intellectual Development of Europe* (1875), wrote in the same strain. A more recent writer than Scott, Robert Briffault, in *The Making of Humanity* (1919), summed up the relationship of the Arabic sciences with those of Europe:

> 'The debt of our science to that of the Arabs does not exist in startling discoveries, or revolutionary theories. Science owes a good deal more to Arab culture; it owes its very existence.
>
> What we call 'science' arose in Europe as a spirit of inquiry, new methods of investigation, the method of experiment, observation, measurement, and the development of mathematics in a form unknown to the Greeks.
>
> That spirit and these methods were introduced into the European world by the Arabs.'

One need not, therefore, be surprised to learn that these same Arabs produced in the homeland of Iraq, in Spain, and elsewhere, a system of farming capable of supporting their brilliant civilization. Many of the books quoted in Scott's list bear testimony to their farming art. I will confine myself, however, to quoting from two well-known French authors in his list. One is from Gustav le Bon's *La Civilisation des Arabes* (1884):

> 'The Arabs had even a greater aptitude for agriculture than for letters and arts. What means of irrigation are now found in Andalusia were made by them.'

The other is from Sédillot's *Histoire Générale des Arabes*:

> 'In short, they had irrigated and cultivated the land so excellently that it was befitting to call Andalusia a garden.'

Martin Hume, writing three years before Scott, summarized the farming art of the Spanish Arabs:

> 'Agriculture and horticulture were developed to an extent never heard of before.'

Scott also lists in his bibliography works on Spanish farming. One of these works that escaped the attempted total destruction of the literature of the Arabs by their fanatical conquerors is *The Book of Farming*, by Ibn-Al-Awam (or to give him his full Arab name, Abu Zackaria Yahya Bin Mohammed Bin Ahmed Ibn Awam), who lived in Seville in the sixth century of the Mohammedan era. Scott includes him in his list with the French translation of his work, *Le Livre de l'Agriculture* (2 vols., Paris, 1866). This book was also translated into Spanish in 1802, and into Urdu in 1927. It was not translated into English during Scott's lifetime, nor, as far as I know, has this omission in English scholarship yet been corrected. The Arabic manuscript, however, is held in the British Museum Library, as well as in the libraries of Leyden, Paris and the Escorial.

Ibn-Al-Awam also has his own list of 107 authorities on the varied aspects of farming, and, since the Arabs were great translators, he quotes freely not only from Arabic writers, but from Greek, Latin, Persian, Nabathean and other agricultural experts, as well as writers on such related subjects as botany, zoology, chemistry, mechanics and meteorology. His translator into Urdu emphasizes that Ibn-Al-Awam was a very cautious student; a true scientist, in short. The translator writes:

> 'The peculiar quality of this book is that, whenever the author quotes the statement of an expert, he first tests it by personal experiment. Where he had no opportunity to verify a statement by experimentation, he tells his readers that, though he has been unable to do so, he has such faith in the veracity of his informant that he has copied his statements into his book. This precaution, which is absent in other books, has greatly increased the value of the work of Ibn-Al-Awam.'

A very reliable man, then, is this Spanish-Arabic scholar. Scott himself describes *The Book of Farming*:

> 'The great work of Ibn-Al-Awam of Seville, a vast monument of industry and erudition embracing every conceivable branch of the subject, shows to what extraordinary perfection the science of agriculture had been carried in the twelfth century by the Spanish Mohammedans.
>
> It covers, in a comprehensive and exhaustive manner, not only the methods found by the experience of centuries to be the best adapted to the sowing and harvesting of grain, the planting and cultivation of orchards, the propagation of edible and aromatic plants; but it also, with infinite detail, describes the breeding and care of every species of domestic animals, their qualities, their relative excellence, their defects, their habits, and their diseases.

It discourses at length upon the different breeds of horses and upon the rearing of that useful animal so prized by the Arab. It explains the details of artificial incubation, a process borrowed from Egypt.

It explains how to produce in geese the abnormal hepatic conditions which induce the *foie gras*, that artificial delicacy so dear to the epicure, and a thousand years ago, as today, an invaluable adjunct to fashionable gluttony.

It teaches the different methods of cooking and the preparation of various confections, jellies, syrups and sweetmeats of every description.

The manufacture of wine, so rigidly forbidden to the Muslim, and whose immense consumption had already, in the time of the Khalifate, scandalized the pious, is detailed in all its stages in this remarkable book.

In it are given recipes for cordials of many kinds, cooling beverages and hydromel.

It also prescribes the rules by which the household of the farmer should be governed, and defines the reciprocal duties of employer and employee.

In every operation of rural life and domestic economy, it enforces by repeated admonition the necessity for cleanliness, system and order.'

I have dealt at some length with the credibility of Scott's account of the Arabic agricultural system in Spain, because, although in the industrial era, which began some 170 years ago, we have made vast strides in the sciences and have far outstripped their initiators (the Arabs), we have at the same time not advanced – but dangerously receded – in the recognition that our own civilization must for our safety and prosperity be founded upon the soil and its preservation.

The agricultural system of the Moors in Spain was, writes Scott:

> 'the most complex, the most scientific, the most perfect, ever devised by the ingenuity of man. Its principles were derived from the extreme Orient, from the plains of Mesopotamia, and from the valley of the Nile – those gardens of the ancient world where, centuries before the dawn of modern history, the cultivation of the earth had been carried to a state of extraordinary excellence.
>
> To the knowledge thus appropriated were added the results of investigation and experiment; from the introduction of foreign plants, from the adoption of fertilizing substances, and from the close and intelligent observation of geographical distribution and climatic influence.'

No cultivators had a more profound knowledge of the value of water than this people. They, like the great riverine peoples from whom they derived so much knowledge, realized that the proper use of water *was* civilization. Without its just and conservative distribution, the true justice and magnanimity of civilization do not really exist. By the art of distributing water,

> '...a considerable portion of the country which had never been subjected to tillage because of its aridity became suddenly metamorphosed, as if by the wand of an enchanter.
>
> Barren valleys were transformed into flourishing orchards of olives, oranges, figs and pomegranates. Rocky slopes were covered with verdant terraces. In districts where no water had even existed, there now flowed streams and broad canals. Where marshes existed, the rich lands they concealed were drained, reclaimed

and placed under thorough cultivation.

On all sides were visible the works of the hydraulic engineer – the reservoir, the well, the sluice, the tunnel, the siphon, the aqueduct. The necessary water was supplied to the fields by every device then known to human skill.'

Water was lifted to higher levels by Persian wheels, of which in a few square leagues there might be 500, some with diameters of 70 feet. Grades were calculated by the use of the astrolabe.

'The public works constructed for irrigating purposes were gigantic in scale.

The artificial basin near Alicante, elliptical in shape, is three miles in circumference and 50 feet deep; the dam at Elche is 264 feet long, 52 feet high, and 150 feet wide at the bottom; that over the Segura, near Murcia, is 760 feet long and 36 feet in height. The aqueduct at Manesis, in Valencia, is 720 feet long, and is supported by 28 arches.

The principle of the siphon, familiar to the Arabs 800 years before it was known in France, was utilized to a remarkable degree in the Moorish hydraulic system. The length of the curve in the great siphon at Almonora is 570 feet; the diameter of the latter is six feet, and it passes 90 feet under the bed of a mountain stream.

The subterranean aqueduct at Maravilla, which waters the plain of Urgel, is a mile long and 30 feet in diameter; that of Crevillenta, north of Orkuela, is 5,565 feet long and 36 feet in diameter.

All of these underground conduits are cut through solid rock. The masonry of the reservoirs is of the finest description, and the cement used has become harder than stone

> itself. Contingencies are provided for, with some skill and foresight, so that no overflow occurs, and no damage ever results, even during the greatest inundations.
>
> The excellence of construction of these massive works of Arab engineering is demonstrated by the fact that they have needed practically no repairs in a thousand years.'

The distribution of the water was governed by a code of laws, perfect familiarity with which was only to be obtained by those working for their livelihood under its direction. With a wise trust in local government, the execution of these laws was presided over by a Tribunal of the Waters, the members of which were chosen by the farmers themselves. This Tribunal saw that there was no waste. Theft was heavily punished, and disputes and violations of the regulations came under its jurisdiction.

> 'Judgment was rendered after consultation, and from it there was no appeal. The most exalted rank, the greatest wealth, the most distinguished public service, did not confer exemption from the jurisdiction of the court or affect the impartiality of its decrees. The noble was summoned to its bar with little more ceremony than the slave... The wisdom of these regulations is demonstrated by their longevity.
>
> In the distribution of water the measurement was by volume, a certain quantity being allotted to a stated area during a given period of the day or night at intervals of ten to fifteen days. The sides of the canals were provided with flood gates, kept under lock and key, by which the adjoining fields could be submerged at the proper time. Drains carried the surplus back into the original channels, so that there was the least possible loss.'

Such was the way in which water was used so as to make a great society of people possible and durable. It is of profound significance, but it is seldom known by modern scholars of even the widest education. There is no knowledge more desperately needed by modern Europe. What is the use of a vast treasury of knowledge which is lacking this vital knowledge which the Arabs possessed?

In the second great precept of the art of agriculture – the rule of return – the Arabs were as effective as they were in the knowledge of water. The same care and economy were observed in fertilizing the soil, which the requirements of a dense population never permitted to rest. Scott writes:

> 'Manure and dust were collected from the highways. The contents of sewers and vaults were preserved, desiccated, and, mingled with less powerful substances, were used to atone for the impairment of the soil caused by incessant cultivation. Ashes, the burned and pulverized seeds of fruits, the blood and bones of slaughtered animals, all played an important part in the intelligent and systematic treatment of the rich and productive valleys of the south, whose surface, resting on an impenetrable subsoil of clay, required continued renovation. The curious and minute investigations of the skilled agriculturist had determined the best composts, the most advantageous modes of applying them, and the kind of vegetation to which they were especially adapted.
>
> Manures were deposited in stone reservoirs contrived to prevent evaporation or leakage. Nothing was wasted; every substance available for the fertilization of crops was carefully preserved, the different varieties being separated and applied to such soils as experience had taught were most productive under their use.'

The third great precept of the art of agriculture was followed by the Arabs, in their preference for independent small holdings.

> 'Unlike the policy adopted under the Roman and Gothic systems, there were few large estates. The land was divided into small tracts, and for that reason was much more thoroughly tilled.
>
> Every indulgence and encouragement to the cultivator was afforded by Moorish law. He enjoyed to the utmost degree compatible with the social order the independence so necessary to the successful prosecution of agricultural pursuits.
>
> For the most part, the farmer himself instituted the regulations of husbandry, which were enforced by magistrates taken from his class and of his own selection. His taxes were not oppressive. The productiveness of the soil and the equability of the climate never permitted his labours to go unrewarded.'

A fourth was the use of terraced cultivation:

> 'In localities unfavourable to cultivation, the deficiencies of the soil were supplied by untiring industry. Walls of masonry supported terraces where the very cliffs were made productive, and where only a bush or vine could be planted, the narrow space was utilized. Not only water, but loam and fertilizing materials were brought from great distances.'

Cultivators were also encouraged to extend their knowledge.

> 'The unrivalled excellence of the agricultural methods employed by the Spanish Mohammedans was, in large measure, due to their profound botanical knowledge.'

Botanists were dispatched to Egypt, Mesopotamia, India, the East, indeed every quarter of the globe, to collect the seeds of useful plants and fruits for experimental cultivation:

> 'Gardens for the propagation of both native plants and exotics were established in the environs of all the great cities, and the results of intelligent observers were regularly tabulated for the public benefit... In all the various duties of his occupation, the Moorish horticulturist possessed expert knowledge.'

As a result of the insatiable thirst of the Moorish naturalists for research, the Arabs introduced into Europe the strawberry, lemon, date, quince, fig, mulberry, banana, pistachio, almond, rice, sesame, buckwheat, spinach, asparagus, mace, nutmeg, pepper, caper, saffron, coffee, cotton and sugar cane, though according to Dr Hintze, in his book *Geographie und Geschichte der Ernährung*, some, such as lemons, quince, almonds and mulberries, had appeared on the tables, if not in the fields, of the Roman Empire.

Botanical knowledge and widespread education, shortly to be described, therefore combined to promote these excellent results. Moorish treatises on agriculture and horticulture dealt with every aspect of cultivation. The cultivators were familiar with the movement of the sap, the difference of sex in plants, and the process of artificial fertilization. They described in plants the conditions of activity and repose, of motion and sleep. They had no less than eight methods of grafting and protected the grafts from the injurious effects of the sun with ingenious devices. They knew how to preserve fruits and grains in subterranean chambers hewn out of the rock. In all agricultural matters, in brief, knowledge was strengthened and widened by skilled agricultural literature.

They possessed the same skill and knowledge in the rearing of cattle and horses, in the breeding of sheep and the culture of bees, all of which attained the highest degree of proficiency. The Arab horse lost none of its speed and endurance for being bred and reared in Spain. The abundant, silky fleece of the merino sheep was due to a peculiar method by which flocks were tended. Immense flocks were driven twice a year between the slopes of the Pyrenees and the plains of Estremadura, by which means they secured both fresh and continual pasture and freedom from the droughts of summer and the storms of winter.

Lastly, the love of flowers was a passion among the Spanish Muslims. Scott writes:

> 'As they were the greatest botanists in the world, so no other nation approached them in the perfection of their floriculture, and the ardour with which they pursued it.'

Whether they were cultivated solely for their beauty and perfume or whether they also cultivated them, because, as has been seen in Chapter 16, they help in special ways to preserve the cultivation of the soil, Scott does not say.

With these fine farming practices, the food of the people was provided, but because of the uncertainties of the weather and in spite of the great system of irrigation, bad years would nevertheless occur. To protect the people against hunger at such times, the export of grain was forbidden – as laid down in the Koran – and the surplus of good harvests was stored in granaries hewn in the rock. Forests of oak were also carefully preserved for the sake of their acorns, which seved as a coarse but nutritious diet when famine threatened.

Such a wide-ranging and complete agricultural system as the Arabs of Spain created confers a wholeness and health to the other aspects of civilization. Consequently, in every other art, there occurred the same prosperity and excellence as those which distinguished the art of the soil.

Of their other arts and industries, Scott writes with a fervour no less inspired by their brilliance than it was by the cultivation of the soil. He describes the organization of the traffic of commerce by land and by sea; the markets and fairs; the principles of equitable dealing in business transactions and in dealing with other nations, as laid down by Islamic law; the ports and the great centres of manufacturing and mercantile activity situated on the Mediterranean Sea; the silk factories and the factories of iron and copper utensils of Almeria; the potteries of Andalusia; the leather work of Cordova, the capital; the silks of Seville; the paper of Xativa; the steel of Toledo; the textile fabrics of Lusitania and Andalusia; the glass-work at Almeria, which informed the later glass-work in Venice; the jewellers of Granada; the mats and basket work of Alicante; the mills of Murcia and Saragossa; the linens of Salamanca; the musical instruments of Seville; and the wines, the use of which scandalized the orthodox Muslim, to whom intoxicants of any kind were forbidden.

Above all these accomplishments of labour was the passion for literature and knowledge. The great monarchs of the period from 755 A.D., when Abd-al-Rahman I founded the Ommeyade Dynasty in Spain, to the death of Al-Hakem II in 976 A.D., were not only enthusiastic patrons of literature, but were themselves personally distinguished as authors.

Abd-al-Rahman I, amidst a life of inexhaustible adventure, from prince to beggared outcast and from outcast eventually to king, was a real lover of literature and art, and a poet of unusual ability. He cultivated public taste by holding literary contests, and attracted the most accomplished scholars and poets to his side, not only by material rewards, but through his friendship and the engaging versatility of his comprehensive genius. Had Leonardo da Vinci lived in his time, he would have found the royal friend, worthy of his genius, whom he sought for in vain in Italy and France.

The successors of Abd-al-Rahman I were worthy; indeed, one can hardly believe how there came into being such a series of monarchs, not just highly educated, but also possessing that high degree of culture which alone can be promoted and nourished by an inward passion for it. Such men have filled thrones in many lands with great benefit to their peoples, but the Ommeyade Dynasty in Europe was certainly unique in the number of its monarchs of high culture. It reached its peak in the reigns of the monarchs regarded as the greatest of the Arab kings of Spain, Abd-al-Rahman III and his son Al-Hakem II, the monarch who exemplified the highest personal culture possibly ever reached by a monarch. Scott says:

> 'The prominent features of the character of Al-Hakem were his love of learning, his profuse but always judicious liberality, and his profound reverence for the doctrines of the Koran and the laws of the Empire. The few military operations he was called upon to direct showed no want of vigour, and suggested that in a less peaceful age he might have obtained the laurels of a successful general. His devotion to literature amounted to a passion. No monarch of whom history makes mention has equalled him in the extent of his knowledge or the number and diversity of his literary accomplishments.'

Al-Hakem gathered together an unequalled library, which required 44 volumes for the catalogue alone.

> 'With the contents of most of these works Al-Hakem is said to have been familiar, and,

> indeed, many of them were enriched by notes and comments written by his own hand. The title page of each volume bore not only the name of the author, but also his genealogy, as well as the date of his birth and death, all collected and preserved by the indefatigable industry of the royal scholar.'

His prodigious memory, his powers of acquisition, his critical acumen, his talent for composition, and the capacity which could abstract from the administration of public affairs of a great monarchy sufficient time for literary undertakings – that Al-Hakem possessed these qualities, which under ordinary circumstances could only be accomplished in a lifetime of constant study, is marvellous and incredible.

Al-Hakem was an historian of approved merit, as well as an impartial critic and a voluminous commentator. He wrote a history of Spain, now unhappily lost, which was considered a work of great authority in its time, and whose reputation was universally held to be independent of the prestige which it would naturally derive from the name and rank of its author. Such was his erudition that in knowledge on obscure points of genealogy and biography he was without rival, even in the learned court of Cordova; and his fund of historical information was so profound, and his judgment so accurate, that his opinions were respected and unquestioned by the most accomplished scholars of the Muslim world.

A prodigious impulse was given to education by this extraordinary patronage of letters. The accumulated wisdom of Africa, Asia and Europe was gathered together at Cordova. Education was provided systematically, with regulations that were enforced with military precision.

Linguists exhausted every source of knowledge. Not only did they translate the masterpieces of Greek and

Roman literature, but they also familiarized themselves with Persian, Chaldaic, Hebrew, Chinese, Hindu and Sanscrit works. This education and

> '...the absolute intellectual liberty which existed was, indeed, considered a reproach by ignorant Muslims of less enlightened lands, who could not understand the association with heretics and the toleration of infidels; but in Spain, where a system of universal education had been established, and was enforced by law as well as by the influence of public opinion, this inestimable privilege was thoroughly appreciated.'

Encouraged by the patronage of royalty,

> '...the mental development of the masses advanced with gigantic strides. In Cordova alone there were 800 public schools frequented alike by Muslims, Christians and Jews... There was not a village within the limits of the Empire where the blessing of education could not be enjoyed by the most indigent peasant.'

Women joined in this advance:

> 'The exalted position occupied by women under the Arab domination in Spain gave them an influence and invested them with an importance unknown elsewhere in the Mohammedan world.'

Chemists, botanists, biologists, astronomers, mathematicians, physicians and surgeons lifted science to a level it had never previously reached in Europe. Engineers covered the land with roads, canals and public works, and architects brought into being the exquisite buildings, the palaces, colleges and mosques which the religious fanaticism of the Christian conquerors later destroyed, together with the libraries and their books.

The farmers and their families had a full share in the education of this great period. In all the principal towns there were schools of agriculture. From them, cultivators learnt to preserve fruits and to protect their fields against noxious insects. They learnt meteorology and could foresee atmospheric changes with accuracy. In all the duties of farming, they possessed expert knowledge. It was in this period that

> '...agriculture was brought to such excellence as seemed to make any further improvement impossible'.

An important measure of Arabic Spain is the extent of the population.

> 'It has been estimated by competent authorities that the subjects of Abd-al-Rahman III numbered at least 30 million. Great as was the extent of the metropolis, incredible as was her wealth, superb as were her environs, many of the other cities of the Empire, while they could not rival her power and grandeur, shared the enormously profitable benefits of a civilization in which Cordova enjoyed a well-deserved pre-eminence.
>
> The dominions of the Khalif included 80 municipalities of the first rank and 300 of the second. The smaller towns were innumerable; along the banks of the Guadalquivir alone stood 12,000 villages.
>
> So thickly was the country settled that the traveller usually passed, in the space of a single day's journey, no less than three large cities in the midst of an unbroken succession of towns and hamlets. Nothing comparable with the opulence and splendour of the great provincial capitals was to be seen outside the Peninsula.
>
> Seville contained 500,000 inhabitants; Almeria an equal number; Granada 425,000; Malaga 300,000; Valencia 250,000; Toledo 200,000.'

The effect of the final expulsion in 1609 of the *Moriscoes* (Muslims who remained in Spain after the Christian conquest and were compelled to become converts to Christianity) is described by Buckle, in his classic *History of Civilization in England* (1861):

> 'The effects upon the material prosperity of Spain may be stated in a few words.
>
> From nearly every part of the country, large bodies of industrious agriculturists and expert artificers were suddenly withdrawn. The best systems of husbandry then known were practised by the Moriscoes, who tilled with indefatigable labour. The cultivation of rice, cotton and sugar, and the manufacture of silk and paper were almost confined to them. By their expulsion, all this was destroyed at a blow, and most of it was destroyed forever.
>
> The Spanish Christians considered such pursuits beneath their dignity. In their judgment, war and religion were the only two vocations worthy of being followed.
>
> When, therefore,the Moriscoes were thrust out of Spain, there was no one to fill their place; arts and industries either degenerated, or were entirely lost, and immense areas of arable land went uncultivated. Some of the richest parts of Valencia and Granada were so neglected that the ability to feed even the scanty population which remained there was lost. Whole districts were suddenly deserted, and down to the present day have never been repopulated.'

The population of Madrid, continues Buckle, fell from some 400,000 to 200,000; Seville's population decreased by three-quarters and her 16,000 looms dwindled to under 300; Toledo witnessed the disappearance of her silk manufactory, which employed 40,000 people, and upwards of 50 woollen manufactories shrank to 13; Burgos became deserted and lost everything but its name. In Buckle's grim words:

> 'Spain, numbed into a death-like torpor, spell-bound and entranced by the accursed superstitions which preyed on her strength, presented to Europe an example of constant decay'.

22

An Historical Reconstruction

THE ISLAMIC CIVILIZATION in Spain was responsible for what was, perhaps, the most remarkable reconstruction of mankind in history. An outline of it can best be given if it is divided into three periods – its initiation, institution, and achievements.

The Initiation

The period of initiation was that of the life of its founder, the prophet Mohammed. Mohammed was born in Mecca in 570 A.D. as a member of the leading tribe of Mecca, the Koraish. His father dying before he was born and his mother when he was six, he came under the tutelage of his grandfather, Abd-al-Muttalib. His grandfather died when he was thirteen, and he was then raised in the family of a poor but affectionate uncle, Abu Talib.

Mohammed grew up to be a quiet, meditative man, taking little or no part in public affairs, but his humanity and sense of justice earned him the name of 'The Trustworthy'.

Then, at the age of over 40, during a night of meditation he heard a voice commanding him: 'Cry, in the name of the Lord'. He obeyed, and henceforth became a messenger, spreading the word of Allah. In Mecca he denounced the idolatry which was the religion of the people at the time, and in its place taught the worship of Allah, the one and only God.

This aroused the fury of the Koraish against him and his

disciples. His followers escaped to Yathreb, or Medina, the City of the Prophet, as it later became named. Mohammed remained behind amongst his enemies. Discovering a plan to murder him, he finally also fled from Mecca to Medina. The *Hegira*, or Flight, took place in 622 A.D., and it is on this date that the Mohammedan calendar commences.

In Medina, the religion he taught was simple. He preached that there was but one God, the unity of living things, the brotherhood of man, kindness to women and children, gentleness to animals, alms for the poor, and the value of prayer.

His preaching and personality won the hearts of the people of Medina. He was made the Chief Magistrate, a post which gave him the power to put into practice what he taught. At that time, writes Ameer Ali Syed, in *A Short History of the Saracens* (1900),

> '...there was no law or order in any city of Arabia.'

Medina at the time was torn by a feud between two principal tribes. Mohammed reconciled the two tribes, abolished all tribal distinctions, and grouped the inhabitants of Medina under one generic name; *Ansar*, or 'Helpers'. He issued a charter by which all blood feud was abolished, and lawlessness repressed. Equal rights were granted to the Jews, of whom there were many in and about Medina, and who, for their part, committed themselves to helping the Muslims in defending the city if it was attacked.

The next step in his mission was to unite the peoples of Arabia, but in this he was hindered by the bitter enmity of his own tribe, the Koraish of Mecca, who in the first year of the *Hegira*, attacked the Muslims and were defeated.

In the third year, the Meccans, under the command of Abu Sufian, the son of Ommeya, whose descendants were to become the Ommeyade Caliphs of the early Arabic Empire, were successful, but their losses were so great that they did not dare to attack Medina itself.

In the fifth year of the *Hegira*, the Meccans besieged Medina with an army of 10,000. In spite of the treachery of the Jews, who took the side of the Meccans in the siege, the Muslims, thanks to Mohammed's defensive skills, were victorious.

This victory gave Mohammed the freedom to extend his work. The prestige which he gained, both as teacher and as general, led to his acknowledgement by tribe after tribe throughout Arabia.

In the seventh year of the *Hegira*, the Meccans attacked one of these tribes. Mohammed gathered together an army of 10,000 men and entered Mecca as a conqueror. Nevertheless, at the sight of the city and of familiar but hostile faces, he treated the Meccans as brothers. Apart from four criminals, all were forgiven and accepted Islam. Mohammed himself shattered the idols of Mecca with the cry:

> 'Truth is come, darkness departs!'

The ninth year of the *Hegira*, known as the Year of Deputations, witnessed the general acceptance of Islam by the tribes of Arabia. Mohammed dealt with them in the same liberal spirit as he had shown to the Meccans. Ameer Ali writes:

> 'A written treaty guaranteeing the privileges of the tribe was often granted, but in order to promote the change of heart that was Mohammed's special mission, a teacher invariably accompanied the departing guests to instruct the newly converted people in the duties of Islam, and to see that every evil practice was obliterated from their midst.'

Mohammed had now fulfilled his mission by uniting all Arabia. He died on 8 June 632 A.D., at the age of 62.

The actions, character and teaching of Mohammed

made so profound an impression upon his contemporaries that his teachings went on to form the basis of Islamic law and civilization. The chief source of that law was the Koran, but the Koran did not cover all the growing needs of the Arabic Empire that followed after the death of Mohammed. Hence, in addition to the Koran, every detail that could be recalled by contemporaries and especially by those closest to him and with whom it was his habit to consult was carefully recorded.

In Medina, he had been the final judge; the spiritual leader of religion, as well as the temporal leader, with all the duties and powers of an essential sovereign conferred on him by the people, by way of their complete faith in 'The Trustyworthy'.

Hence, in the formation of Islamic law, his inspired utterances in the Koran, his discourses, his decisions after consultations, his expressed approvals, his tacit agreements by gesture, his judgments, his actions – were all brought into service as guides of conduct for pious Muslims by the Islamic legalists and the Caliphs. From them were derived the fundamental or fixed laws; fundamental and fixed because they were derived from the Prophet of God.

Islamic law controlled every aspect of the life of a Muslim, but for simplification of this vast subject it was divided into categories, ranging from the few obligatory or prohibited things as determined by the Koran and the *Hadith* (or traditions) of the Prophet, to the limited number of the approved or disliked, and the unlimited number, which were left to everybody's common sense.

Bylaws were set in place to adjust to changes in circumstances which time brought about, but they were kept within the orbit of the fundamental laws. Only in cases of extreme emergency could a fundamental law be abrogated, and then only for the duration of the emergency.

Never, therefore, in history has any man been so intimately identified with a civilization as was Mohammed with that of Islam, which endured as an empire until the sacking of Baghdad by the Tartars six and a quarter centuries after the death of Mohammed. This civilization produced a truly remarkable reconstruction of mankind in agriculture, manufacture, trade, knowledge, art, and all the other aspects of human society.

The spirit of Mohammed's precepts become peculiarly important at a time such as the present, when a further reconstruction is so urgently needed.

This spirit has been admirably laid out for readers of English by Ameer Ali Syed in *The Spirit of Islam* (1922). Its chief character was Mohammed; he was the first statesman to introduce decency of human conduct into every part of society. He left no class of human beings out. He was, in short, one of the greatest humane and constructive statesmen in history.

Ameer Ali, however, does not give an explicit account of Mohammed's attitude to war, which is of such vital concern to men in these days. So, before taking up his review, it we will first briefly consider Mohammed's ideas regarding war. They have been decscribed by Marmaduke Pickthall, in an article entitled *War and Religion* in the *Islamic Review.*

Firstly, Mohammed recognized the inevitability of war in the collective life of mankind;

> 'If it had not been for Allah's repelling of some men by others, the world would have gone to badness; but Allah is a lord of kindness in creation.' (the Koran)

To repel bad men was, therefore, the reason for going to war and it was for this reason that every capable Muslim must be prepared to go to war if called upon to do so. It was not conscription, but a sacred duty, provided that the war was a holy war, or *Jihad*. The Koran says:

> 'Fighting is enjoined upon you, and it is a thing hateful to you. But it may be that you hate a thing which is good for you, and it may be that you love a thing which is bad for you; God knows best and you do not know.'

The Koran enjoins war against grave injustice:

> '...to defend the weak man, and women and children, and those who say: "Our Lord, take us out of this city whose people are oppressors. Oh, send us from Thy presence a befriender; oh, send us one who can help us!"'

Retaliation against aggressors was commanded:

> 'Kill them wherever you find them and drive them out of the places from which they drove you out. Persecution is more cruel than killing. And do not fight them round the sacred mosque, unless they attack you there. And if they do attack you, kill them. Such is the reward of graceless people.'

But on no account were the Muslims to be the aggressors.

> 'Fight in the way of Allah against those who fight against you, but do not originate hostility. Truly, Allah loves not the aggressor.'

By the spread of Islam, therefore, Mohammed hoped to abolish the brutality – or even existence – of war. Throughout the Koran, writes Pickthall,

> 'the word "treaty" means a sacred compact, a solemn covenant, which to break is impious.
>
> With Islamic nations, treaties have always had this sacred character. I cannot recall a single instance of a Muslim power ever consciously breaking a treaty, though they have the right to throw the treaty back if they fear treachery.'

Actual treachery was to be treated with the severest punishment, such as was inflicted upon the Jewish traitors of Medina.

Lastly, Muslim soldiers were forced to observe correct or decent conduct. The sanctity of the soil was to be respected. Muslims invading a country were forbidden to destroy fields of corn, or palms, or any fruit trees, or to slaughter cattle except in case of urgent need. Mohammed commanded them to

> '...destroy not the means of subsistence.'

Similarly 'the quiet people', as the Muslim jurists called the unarmed inhabitants, were to be respected. They were not to be killed; they were not even to be molested; neither they nor their houses were to be plundered. As the Prophet said:

> 'Plunder is no better than carrion.'

Material left on the field of battle was lawful booty, however. Finally, enemy combatants were to be respected:

> 'If they desist (from fighting), then (there should be) no hostility, except to evil-doers.' (the Koran)

For the evil-doers, though, there was the law of retaliation. As they had done, so should it be done to them.

Now, under the guidance of Ameer Ali, we will review Mohammed's attitude to conquest. In this no statesman ever used more effectively the quality of clemency to those forced to acknowledge his authority.

To those who accepted Islam, he ordained all the privileges and freedom associated with the name, meaning as it does 'surrender to Allah'. To those who submitted, but wished to keep their own faith (other than that of idolatry), he presented the utmost tolerance. They were allowed to pursue their own customs and their religious faith, provided they paid their taxes and obeyed the other civic duties imposed on them by their Arabic rulers. They were exempted from military service, paying a special tax in lieu of it. Their lands were not taken from them.

The precedent of this tolerance was set by the Charter which the Prophet granted to all Christians in the sixth year of the *Hegira*. The spirit of it was Christian in its best sense, since Mohammed always regarded Jesus as the teacher most akin to him in time and teaching. Ameer Ali writes in his *History of the Saracens* (1900):

> 'In this Charter, the Prophet enjoined his followers to protect Christians, to guard them from all injuries, and to defend their churches and the residences of their priests.
>
> They were not to be unfairly taxed; no bishop was to be driven out of his bishopric; no Christian was to be forced to reject his religion; no monk was to be expelled from his monastery; no pilgrim was to be detained from his pilgrimage; nor were the Christian churches to be pulled down for the sake of building mosques or houses for Muslims.
>
> Christian women married to Muslims were to enjoy their own religion and not be subjected to compulsion or annoyance of any kind on that account. If Christians should need assistance for the repair of their churches or monasteries, or any other matter pertaining to their religion, the Muslims were to assist them.'

In pre-Muslim Arabia, women were the chattels of the men. Ameer Ali, in *The Spirit of Islam*, writes:

> 'In both the Persian and Byzantine Empires, women occupied a very low position in the social scale. Fanatical enthusiasts, whom Christendom in later time canonized as saints, preached against them and denounced their enormities.'

Then, when the family, and with it the whole social fabric, was falling to pieces on all sides, Mohammed introduced his reforms and

> 'enforced, as one of the essential teachings of his creed, respect for women.'

Mohammed raised women to legal and economic equality with men. His precepts and the fixed laws on divorce were strikingly just to women, though he himself expressed his strong disapproval of divorce, in that it brought evil and hardships upon children.

So also, as regards property, the rights which he gave to women, in spite of the later deterioration of their status under Persian and Byzantine influence, were such as even now have not been fully attained in most Western countries. Mohammed's aim was to enable women to become independent individuals in the State, and this independence he achieved by allowing them to own property, to possess that which they earned by their own efforts, to have their share in the inheritances left by their fathers, husbands and other near kinsfolk, to be given marriage settlements from their prospective husbands in their favour, and to possess the right to act in any legal matters concerned with these rights without any intervention on the part of their fathers or their husbands. To the best of his power – and his power was great in spite of the opposition of the times – he was an emancipator of women.

Following his precept of the brotherhood of men, Mohammed strove also for the betterment of the slaves. Slaves formed a large part of every society of the time. Ameer Ali describes the Christian attitude to slavery:

> 'The Church itself held slaves, and recognized in explicit terms the lawfulness of this baneful institution.'

Though Mohammed himself abhorred slavery and taught that no action was more acceptable to Allah than the freeing of a slave, he did not attempt the total abolition of a custom so deeply rooted in the economic life of society. What he *did* do, however, was to infuse the whole question with the spirit of brotherhood, and in doing so he altered entirely the character of slavery.

He provided funds out of the public treasury to enable slaves to purchase their freedom without interference from their masters. They could purchase their liberty with the wages of their service. In many ways, he opened up the path of liberty. He ordained decency of conduct to slaves, who were to be treated by their owners with the same kindness that they showed to kindred and neighbours. The slave mother was not to be separated from her child, nor the father from the son, or the husband from the wife. There was to be equality of food between slaves and their owners, and equality of dress. They were only to be addressed in terms of affection and not with words implying a degraded position. Ameer Ali says:

> 'The whole tenor of Mohammed's teaching made permanent chattelhood or caste impossible. It is simply an abuse of words to apply the word slavery, in the English sense, to any status known to the legislation of Islam.'

By abolishing all distinctions of race and colour, black and white, citizens and soldiers, subjects and rulers, Mohammed gave an equal humanity to slaves.

> 'In the field or in the guest-chamber, in the tent or in the palace, in the mosque or in the market, they mingled without reserve and without contempt.'

Insofar then as slavery continued, Mohammed made it a social condition within the brotherhood of man. Muslim slaves could rise to high positions in a state. Some were to become kings; others became governors of provinces, generals, and famous men of learning and religion.

Dealing with the most pressing economic difficulty, that of the distribution of wealth so as to avoid the extremes of the very rich and the degraded poor, Mohammed displayed the rarest wisdom of statesmanship. This was evidenced in the *Zakat*, the rules of inheritance, and the abolition of usury. The story of that great economic work has recently been retold by M. Hamidullah, in the second number for 1926 of the quarterly *Islamic Culture*, in his article *Islam's Solution of the Basic Economic Problems*.

Mohammed, in the Koran, frequently declared that it is for God to provide a livelihood to every creature:

> 'We have given you power in the earth and appointed you therein a livelihood.'

It was the duty of the State, by means of the *Zakat*, or growth tax, to ensure this livelihood. *Zakat* was a tax on all property owned beyond a certain maximum and was meant, as Mohammed said,

> '...to be taken from the rich among them, in order to be given to the poor.'

If the treasury was not sufficient to supply the needs of the poor, the ruler could compel the rich to do so. The poor he defined as

> '...they who have not the wherewithal to make themselves independent.'

Zakat was of two kinds; *Sadaqah*, or the tax on the growth of capital goods; and a tithe (or tax) on the surplus produce of the soil. The Koran stated that

> '...the *Zakat* is only for the poor and needy, for those whose hearts are to be reconciled (i.e. those who had become impoverished by accepting Islam).

> ...and to free the captives and debtors, and for the cause of God, and for the wayfarer; a duty imposed by God.'

Hamidullah points out some of the particular virtues of this tax and the balance it effected between rich and poor. It gave the workers a degree of security and so increased their productive efficiency, and it justified the prohibition of begging, stealing, and indolence by the Koran.

As all superfluous wealth was regarded as productive and was, therefore, taxed whether it was put to use or left unused, it prevented employers taking unfair advantage of labourers. If the workers went on strike, the money and property of the employers continued to be taxed. It prevented deliberate or careless hoarding, for any hoard was taxed.

> 'Let not those who hoard up that which God has bestowed upon them of His bounty think that it is better for them. Nay, it is worse for them...,'

the Koran said. Hoarding for the sake of the family was likewise forbidden, for the Koran declared:

> 'Among your wives and your children are enemies for you, therefore beware of them. Let not your wealth nor your children distract you from the remembrance of God. Establish worship and pay the *Zakat*.'

No rich man could be a Muslim without paying the *Zakat*.

Finally the Prophet believed that so great would be the prosperity resulting from a greater equalization of wealth that a time would come when people offer *Sadaqah* and there will be none to take it.

The second of Mohammed's measures to prevent the accumulation of wealth in a few hands lay in the principles of inheritance. Private property could be accumulated in a man's lifetime within due restrictions, but at his death it was widely distributed amongst his offspring and kindred, and thus large individual fortunes were dispersed amongst many individuals.

The third measure was the forbiddance of usury (or interest on money, as the dictionary defines it). Money never had interest attached to it, and so never burdened the sacred duties of farming and trade with debt. Only the original sum of a loan was to be repaid, otherwise the interest on a loan would make it destructive. In one of his most searching and prophetic sayings, Mohammed seized upon this truth:

> 'Although interest brings increase, yet its end tends to scarcity.'

Money was meant to assist trade by the method of partnership. It was not to be hoarded or lent out at interest. It must be used for trade or spent in alms, said the Koran, so that the *Zakat* due on it do not swallow it up.

By means of partnership, the lender or partner took his share of the success or failure of the enterprise.

> 'Trade is just like interest-taking, whereas God permitteth trading and forbiddeth interest.'

Genuine partnerships were encouraged to further trade, manufacture and farming, but debenture-holders and commercial loans were ruled out as destructive. The imposition of the *Zakat* and the prohibition of interest forced money into circulation and into the promotion of the general prosperity which resulted from its use.

Economic ranks and occupations did not affect the general freedom of the individual. Islam removed from money its power as a standard of social distinction. A man was wealthy according to the good he did to others. Monetary wealth had only a limited value, whereas virtue could not be measured other than by the good to mankind that followed from it.

Three taxes were attached to the products of the crust of the earth – the tithe; the *rikaz*, which assigned one-fifth of the products of mines exclusively, like the tithe, to the poor; and the *kharaj*, a levy for the general welfare of about 2.5 per cent on the output of the land, due regardless of whether the owner cultivated the land or not.

According to Islam, land is a gift to all men, and all men are united in their dependence for sustenance upon the soil. Yet not all could own land. So the land was not socialized, but its products were socialized by these taxes. Through them the poor were given their measure of independence, and the general welfare was given an economic basis in the land. As the soil depended upon the use of everything that nourished it, so the soil, in its turn, was made to give nourishment to all, and to produce the social balance that derived from it as an integral factor of the life cycles of man.

Based on the limitations of the soil's productivity, the economics of Islam dictated a limitation to the acquisitiveness of individual men. As an outgrowth of this, Mohammed's instructions on leisure were also directed so that people's attention was diverted to other things than the making of individual fortunes. Through learning, service, and the call to prayer five times a day, people's leisure was directed to self-cultivation, whereas their working hours were directed to the cultivation and distribution of material goods. This was possible, sums up Hamidullah, because Islam was a religion, and not merely an economic system.

With this independence that Islam gave to the individual, labour was elevated as a general duty and both commerce and farming were held to be meritorious in the eyes of the Lord. The cultivation of the soil was regarded by labourers and rulers alike as a sacred duty; Mohammed himself ploughed his own land. The contemptuous sneer which turned the Latin *paganus* or 'villager' into *pagan*, and the man of the heath or field (Anglo-Saxon *haeth* or 'heath', Gothic *haithi* or 'field') into *heathen* was utterly foreign to the sanctity with which Mohammed and Islam endowed the duties of both.

Having freed women from their traditional subordination to men, slaves from their ignominy, the poor from their destitution, and farming and labour from their subordination, Mohammed turned to the liberation of minds from ignorance.

He made education incumbent upon every Muslim, male and female, and sought thereby to influence the minds of all people with the passionate emphasis he laid upon the value of knowledge. Ameer Ali describes this passion in these sayings of the Prophet:

> 'Acquire knowledge, because he who acquires it in the way of the Lord performs an act of piety; who speaks of it, praises God; who seeks it, adores God; who dispenses instruction in it, bestows alms; and who imparts it to its fitting objects, performs an act of devotion to God.
>
> Knowledge enables its possessor to distinguish what is forbidden from what is not; it lights the way to Heaven; it is our friend in the desert, our guide to happiness; it sustains us in misery; it is our ornament in the company of friends; it serves us as an armour against our enemies.
>
> With knowledge, the servant of God rises to the heights of goodness and to a noble position, associates with sovereigns in this world, and attains to the perfection of happiness in the next.'

He would often say:

> 'The ink of the scholar is more holy than the blood of the martyr,'

and he repeatedly impressed on his disciples the necessity of seeking knowledge...

> '...even unto China. He who leaves his home in search of knowledge walks in the path of God. He who travels in search of knowledge – to him God shows the way to paradise.'

Ameer Ali Syed finally gives this summary of the teachings of Mohammed in Medina:

> 'Islam gave to the people a code which, however archaic in its simplicity, was capable of the greatest development in accordance with the progress of material civilization.
>
> It conferred on the State a flexible constitution, based on a just appreciation of human rights and human duty.
>
> It limited taxation, it made men equal in the eye of the law, it consecrated the principles of self-government.
>
> It established a control over the sovereign power by rendering the executive authority subordinate to the law – a law based on religious sanctions and moral obligations.
>
> "The excellence and effectiveness of each of these principles," says Urquhart, "gave value to the rest; and all combined to endow the system which they formed with a force and energy exceeding those of any other political system. Within the lifetime of a man, though in the hands of a population that was wild, ignorant and insignificant, it spread over a greater extent than the dominions of Rome. As long as it retained its primitive character, it was irresistible!"'

With their personal experience of these and other Islamic precepts in action in Medina, the Arab leaders went forth upon their great liberation and reconstruction of the lives of many millions of oppressed people.

The Institution

Before telling the story of the institution of this reconstruction, it is essential to give a brief description of the conditions of the masses, in what Ameer Ali names the 'west' and the 'east'.

His 'east' does not include the great farming country of China, which (it seems that different parts of the world can often be similarly affected at the same time) was engaged in a reconstruction of its own; that of the Tsing Tien system under the Tang Dynasty (618–905 A.D.) after a long period of divided states and Tartar conquests.

Ameer Ali writes:

> 'In the west as in the east, the condition of the masses was so miserable as to defy description. They possessed no civil rights or political privileges; these were the monopoly of the rich and the powerful, or of the priestly classes.
>
> The law was not the same for the weak and the strong, the rich and the poor, the great and the lowly.
>
> In Sassanid Persia, the priests and the landed proprietors, the *dehkans*, enjoyed all the power and influence, and the wealth of the country was centred in their hands. The peasantry, and the poorer classes generally, were ground into the earth under a lawless despotism.
>
> In the Byzantine Empire, the clergy and great magnates, courtesans and other nameless ministrants to the vices of Caesar and proconsul were the happy possessors of wealth, influence and power. The people grovelled in the most abject misery.

> In the barbaric kingdoms – in fact, wherever feudalism had established itself – by far the largest proportion were either serfs or slaves. Villeinage or serfdom was the prevailing status of the peasantry.'

The 30 years of the story, from 11 A.H. to 40 A.H. of the Muslim calendar, were occupied with the settlement of Persia, Iraq, Syria, Palestine and Egypt. Under Omar, the second Caliph, in 21 A.H., occurred the 'Victory of Victories' at Nehawand, in which the Persians, who outnumbered the Arabs by six to one, were totally defeated. Egypt, too, was conquered.

Owing to these victories, the precepts of Mohammed affected the fate of many millions of people. By the fire which Mohammed lit, masses of lowly and oppressed people, as well as people of power and wealth, were warmed and enlightened to a new life.

Convinced that the stability of the Empire and its material development depended upon the prosperity of the agricultural classes, Ameer Ali describes in *A Short History of the Saracens* how Omar

> '...took immediate steps to settle the peasantry securely in their possessions. They were released from the galling oppression of the large landowners; their assessments were revised and placed on a stable basis; the broken aqueducts were restored and new ones built...
>
> Egypt, Syria, Iraq, and Southern Persia were measured field by field, and the assessment fixed on a uniform basis. The record of this magnificent cadastral survey forms a veritable catalogue, which, beside giving the area of the lands, describes in detail the quality of the soil, the nature of the produce, and the character of the holdings.'

The *Zakat* gave independence to the poor, but the rich were not oppressed, even though shorn of their excesses to promote a greater equalization of wealth. There was no communistic division of their lands, nor were they taken by the Arabs. The landholders kept their estates, subject to a fixed tax.

> 'Liberty of conscience was allowed to everyone, and Muslims were ordered not to interfere with the religion of the people. Those who adhered to their old faith received the designation of *Zimmis* (the protected people, or 'liege men').
>
> The sole inducement to proselytism, if inducement it could be called, consisted of the fact that whereas the Muslims, who were liable to be called at any time to serve in the army, contributed only a tithe to the State, the *Zimmis* paid a higher tax in consideration of being exempted from military service.'

Nevertheless this *jazia*, or poll tax, was in no way onerous. When Omar died in 23 A.H., after a reign of ten years, Othman was elected Caliph. Othman was a member of the Ommeyade family of Mecca, of the clan of the Koraish that had shown itself most active in its hatred of Mohammed. The aged Othman was elected to the Caliphate through the intrigues of the rest of the Ommeyade family.

The Ommeyades then got themselves appointed as governors of the provinces.They seized the land and subverted the precepts and actions of Mohammed and the first two Caliphs of the Republic, Abu Bakr (11–13 A.H.) and Omar (13–23 A.H.), who were both early converts and devoted companions of Mohammed. The Ommeyades treated the conquered peoples as satellites and slaves, and enriched themselves by oppression. This aroused the intense hatred of true Muslims, and led to an insurrection in which Othman, at the age of 82, was slain in 34 A.H.

Ali, the beloved adopted son and later son-in-law of Mohammed, was elected as the fourth and last Caliph of the Republic. No man was more revered or trusted by Muslims than he, not only because of his intimate association with the Prophet, but because he was

> '...the truest-hearted and best Muslim of whom history has preserved the remembrance,'

and, because both before and during his Caliphate he so stoutly upheld the doctrines of Mohammed and, during Othman's reign, upheld and extended the practical improvements of Omar. Then, wrote the French historian Sédillot:

> 'One would have thought that all would have bowed before this glory so pure and grand; but it was not to be.'

The Ommeyades were the chief cause of the failure of Ali and it was through their intrigues that he was assassinated after a brief reign (34-40 A.H.). The Ommeyades were supported by tribal chiefs who had been largely weakened in authority by Mohammed's reliance on the faithful Muslims of Medina.

The all-too-human hatred of these reactionaries may seem to have its justification in their deposition from free and arbitrary authority. But their fury and tenacity, of which the modern reader can scarcely form a conception, had their origin in the very roots of the pre-Muslim conditions of the Arabian people. From the very earliest times, blood feud had been bitterly active amongst the tribes, and was consummated in the hatred that existed between the nomadic Arabs of the desert and the farming people of the more fertile south of Arabia, bordered by the Arabian Gulf, in the area known as Yemen.

Reinhart Dozy, in his *Histoire des Mussulmans d'Espagne*, translated into English by G. Stokes (1913), wrote:

> 'This blood feud has endured for 25 centuries; it can be traced back to the earliest historical times, and is far from extinct today.'

To the question of why it preserved its bitterness with such extraordinary tenacity for so many centuries, Dozy wrote:

> 'Handed down from generation to generation in spite of community of language, laws, customs, modes of thought, religion, and, to some extent, of, origin – since both races were Semitic – we can only say that its causes are inexplicable, but that it is "in the blood".'

It may be, as we saw in *Chapter 6*, that the cause of the hostility lay in the ultimate relation of the two peoples to the soil – farming and trade on the one hand and nomadism on the other – and also in the tenacity that characterizes the Semitic peoples. Whatever the true explanation, this passion for feud and tribal independence runs, as a rebellious and anarchical spirit, through the pages of Arab history.

It brought about the failure of Ali's courageous attempt during both Othman's and his own reign to uphold and re-establish the precepts of Mohammed; it led to his assassination by Ommeyade intrigue; it brought the Ommeyades to the leadership of the Muslim world, instigated the persecution of the family of Mohammed and the sack of the sacred city of Medina; it foiled the period of conquest of the Ommeyades and brought about the defeat of the Muslims by the Christians at Tours on the bank of the River Loire; it played its part in the disintegration of the Kingdom of Spain; and it promoted the disintegration of the Saracen Empire of the Abbasides, and caused it to fall to pieces before the assault of the nomadic Mongols. Ameer Ali says:

> 'It led not only to the end of the Republic, but also to the downfall of the Saracen Empire.'

Today, it seems that the Arabs are prevalently nomadic.

The Ommeyade Dynasty which followed the assassination of Ali had its capital at Damascus and endured for 90 years. Only one Caliph of the dynasty, Omar II, strove to re-enact the precepts of Mohammed. He, like Ali, was assassinated. The Ommeyades were themselves finally destroyed by the Abbasides, the descendants of Abbas, the uncle of the Prophet. One, Abd-al-Rahman, escaped to Spain and there founded the great Spanish dynasty of the Ommeyades.

The Achievement

The Abbasides ruled from 750 A.D. to 1258 A.D., when their capital, Baghdad, was seized by the nomadic Mongols. 800,000 inhabitants (*estimates vary widely – ed.*) were butchered within a week, and the great system of irrigation destroyed.

It was under the Abbasides that the great task of reconstruction was accomplished. Mansur, their second Caliph, was the first of a series of brilliant Caliphs, equal to that of the contemporaneous dynasty of the Ommeyades of Spain.

The story of the development of the civilization of the Moors has many resemblances to that of the Abbasides, for both practiced statesmanship based on the fixed Islamic laws. The Abbasides brought the era of conquests to an end. They renounced further warlike enterprises, and devoted themselves to the development of the land, the prosperity of the peasants, the promotion of commerce, the construction of roads and caravanserai, the establishment of charitable institutions, the spread of education, and the encouragement of literature and the arts.

The system of irrigation which the Abbasides developed and extended was one of the most wonderful in the world. Only China and Islamic Spain had anything to compare with it; in fact, the Abbaside irrigation was superior to that of the Chinese, for it had control of the whole of the two great rivers, the Tigris and Euphrates, whereas the Chinese had no control over the sources of their great rivers. The spirit and practice of the great riverine civilization of Babylon were revived.

Throughout the whole Abbaside Empire, the work of promoting agriculture was regarded as a religious duty, and the art of cultivation was developed and maintained with religious zeal.

Mansur first abolished the payment of the Ommeyade money tax upon grains and replaced it with a tax paid in kind. He extended this principle to other crops, and, in the case of the most fertile of lands, the rate of taxation was fixed at two-fifths of the whole. Remission of taxes was frequent in times of stress, even during the reigns of his most severe successors.

By thus following the true economics of the soil, the prosperity of the peasants was set in place, and the soil itself was confirmed as the basis of the State. The method of land taxation was, however, not uniform throughout the Empire. Ameer Ali in his *History* writes:

> 'In Babylonia, Chaldea, Iraq, Mesopotamia and Persia there were numerous landowners and peasant freeholders whose rents were permanently fixed on the basis of agreements entered into at the time of the Conquest. No variation could be made in the tax placed on them, and they were thus protected. The same status was enjoyed by the village communities of Northern Persia and Khorasan.'

In a brief time, under this just system, the countryside of Iraq and Southern Persia took on the appearance of a veritable garden. Between Baghdad and Kufa especially, there were a number of prosperous towns, flourishing villages, fine villas and a teeming population. According to Mohammed Fadhel Jamali, Director of Education for the Ministry of Education in Baghdad, in his book *The New Iraq* (1934), the population was perhaps 500 times what it is today.

A further feature of great value to farming was the principle of self-government – that freedom of local customs and traditions on which Mohammed laid such stress. The Abbasides spread this precept throughout their dominions.

> 'The government carried its policy of non-interference in communities sometimes to the extreme, to the detriment of its own interests. Each village, each town, administered its own affairs, and the government only interfered when disturbances arose, or the taxes were not paid.'

But so vital was the land, and so large the system of irrigation which nourished it, that the construction of new canals and the cleansing and repair of old ones were kept entirely in government hands, as was the maintenance of an efficient river police.

The cost of the new canals was borne by the State, and that of cleansing and repair was shared between the State and the end users of the water. As a result, the workers on the land opened up by the new canals began their tenancies without the burden of a debt that had to be paid off. The benefit to the new farmers was great, and, from their produce they paid taxes to the State, which repaid them with many benefits and services. In this way the soil was dominant, and money secondary.

With the same magnanimity as they bestowed on the soil, the Abbasides developed the precepts of Mohammed regarding education. Academies, colleges and schools were established; education was open to all, urban and rural; the education of women proceeded on lines parallel to that of men. This zeal for knowledge was developed to the highest pitch, as Ameer Ali writes in *The Spirit of Islam*:

> 'Under the Abbasides, we find the Muslims to be the repositories of the knowledge of the world. Every part of the globe is ransacked by the agents of the Caliphs seeking the intellectual wealth of antiquity; whatever they find is brought to the capital, and laid before an admiring and appreciating public.
>
> Schools and academies spring up in every direction; public libraries are established in every city; the great philosophers of the world are studied side by side with the Koran. Galen, Dioscorides, Themistius, Aristotle, Plato, Euclid, Ptolemy and Apollonius; all receive their due appreciation.
>
> The sovereigns assist at literary meetings and philosophical disquisitions. For the first time in the history of humanity, a religious and autocratic government allies itself with philosophy, both preparing and participating in its triumphs.'

What this zeal for knowledge meant for farming, we have already discussed in *Chapter 21*.

This great reconstruction was to be witnessed in every country where Islamic culture was implanted. It was the same story in Persia, Syria, Iraq, Egypt, Mauritania, Sicily and Spain. It seemed as if there was something magical, something beyond all previous conceptions of man, in the arrival of Islam. Spain, Mauritania, Sicily and other countries, previously stagnant or in decay, blossomed into active life.

Idris of Medina, for example, escaped from a false charge of drunkenness. He won the support of the Berbers of Mauritania and founded the Idriside dynasty. He built Fez and made it his capital. Imbued with the new spirit, Fez became a famous seat of learning, and the country of which it was the capital became wealthy and prosperous.

Musa, Abd-al-Rahman and their successors in Spain, Majorca, Minorca, Sardinia, Corsica and Sicily in a very short time established a new culture of prosperity.

There is nothing like this in all history. The early Roman Empire and the modern scientific era are no parallels, because they both progressed, as we have seen, at the expense of the peasantry and the exploitation of the soil.

On the other hand, growth in prosperity in the Islamic countries occurred in *all* branches of social life. Farming, manufacture, trade, art, education, knowledge – all attained a high level. They increased in power and capacity equally. They attained a balance amongst themselves, because they based themselves on a highly developed and conserved life cycle based on the soil, of which the spirit of Islam was the guardian.

23

Summary

IN THIS CHAPTER, designed to sum up the principles of reconstruction by way of the soil, I could have used civilizations other than those of the period of Islamic success as examples against which to measure our own present predicament. O. Cook, of the U.S. Agricultural Department, has provided words which I have already quoted, and I will repeat them here:

> 'Agriculture is not a lost art, but must be reckoned as one of those which reached a remarkable development in the remote past, and afterwards declined.'

This is his conclusion after his examination of the farming system of ancient Peru.

William Prescott, in his *History of the Conquest of Peru* (1847), gave a brief account of the farming practices of the time. The land, he wrote, was divided into three parts: one for the support of the national religion and the sick and infirm; one for the maintenance of the Royal Family and government; and one divided, per capita, in equal shares for the people. By law each man had to marry at a certain age, and the land was reallocated each year, a family's share being increased or diminished according to the number of the members of the family. None were allowed to be idle – from the child of five years, to the aged matron.

Prescott then discussed this agrarian law, and of the European countries that resembled it, he selected Judaea as the one most closely resembling the example of Peru.

However, the information available to me concerning Judaean agriculture is so scanty and remote that, thorough and practical to the highest degree it appears to have been, it would have been quixotic for me to have made it a measure of choice.

The same applies to the riverine civilizations of Iraq; knowledge of essential details is lacking. These civilizations collectively exemplify a stability of the soil over some 4,000 years, but so remote is this period that again, it would not be productive to choose ancient Iraq for modern guidance.

It is a very different matter with a far more numerous people, also possessing a history of 4,000 years. The Chinese, a people whose farming, right up to the years preceding the Great War of 1914–18, earned the unstinted praise of that genius of agriculture, Professor King. The introductory chapter of his *Farmers of Forty Centuries* is a well-deserved paean to the Chinese farming, carried out in spite of the floods of their great rivers, which, forming in the vast area of Tibet, have been beyond their control.

Moreover, they have in their history records of several reconstructions by way of the soil, one of which, that of the Tangs, was contemporary with the Islamic reconstruction. When their society was disintegrated by the incursions and conquests of the Tartars and when the land was devastated, the first task of a Chinese dynasty after overcoming the nomads was the reconstruction of their peasant-based farming system.

In the West, there has been abundant study of the arts of China, and especially its pictures and ceramics, but scholars have largely ignored the greater art; that of agriculture and its reconstruction.

Dr Ping-Hua Lee of Columbia University has been a fortunate exception, however. This gifted author has given accounts of the Han and other dynastic reconstructions of the system of land tenure of the Chinese sages. I made use of his work in the third chapter of my *Restoration of the Peasantries*, under the heading *The First Agricultural Path*.

The history and character of these reconstructions, and of Chinese farming generally, to my mind and to that of Professor King, offer a wide field of invaluable research to future western students, but that time has not yet come. When it does come it will, no doubt, reveal a number of principles of reconstruction by way of the soil at present not available to the West.

For these reasons, therefore, I have chosen the Arabic reconstruction, and also for the further reason that they were, according to many scholars, the initiators of many of our modern sciences. It is true that we have surpassed their sciences to a large degree, but the same cannot be said of their arts, and particularly of their farming as a national art. Here we have by no means surpassed them; on the contrary, we are far below their level.

It is true that through their violent jealousies and extravagances, such as the Chinese were never guilty of, the Arabs exposed themselves to their enemies who destroyed their empire, and though, with a fatality that seems as inexorable as it is inexplicable, they have almost reverted to their original status of desert-dwellers. Even though the Arab cultures today nowhere exhibit any art of agriculture for our enlightenment, I have chosen their historical reconstruction as the measure of what should be possible to us in our present urgent need.

Consequently, to give coherence to my subject before my final chapter, I propose in this chapter to review the first 20 chapters in the light of *Chapters 21* and *22*.

Chapter 1 sets out the general theme and purport of the book; the need in a sane and sound civilization to accept the priority of the soil. It describes the intimacy and oneness of humanity with the soil, and how it forms the basis of the cycles of life in which humanity takes part. It reveals the wholeness and health which arises from a complete adherence to the life cycle, with an account of the Hunza, possibly the healthiest people on the earth. It goes on to show how the human family and the life of the soil are vitally interwoven. It ends with an account of the most enduring association between the soil and the family in history – that of the Tsing-Tien system of the Chinese.

The duration of the Chinese family system and the degree of positive health of the Hunza both surpass what the Arabs attained. Nevertheless, Islam attached great significance to both family and health. Mohammed emphasised the sanctity of agricultural work and taught that the land was inviolable as long as it was rightly used. Islam, founded upon the Prophet's teachings, embodied them in its fixed laws regarding the freedom and security of the peasantry and the inviolability of property when rightly used.

As regards health, we know that, at the time of the rise of Islam and after, Europe was frequently devastated by epidemics. The condition of the towns and the homes of the people was one of extreme filth, and this condition has lasted amongst the poor urban classes almost to the present day. In many of the most populous capitals of Europe not a single public bath was to be found, and religion itself made personal dirtiness almost synonymous with holiness.

The practice of Islam was the very opposite of this. Mohammed himself taught the paramount importance of hygiene. He also placed correct diet as the first source of health, and decreed that lack of restraint in food and drink was the source of all physical ills. Islamic civilization was marked by its insistence on bodily cleanliness, and public baths were provided on a liberal scale. Drainage in towns was efficient without being wasteful.

Though vastly superior in these respects to anything in Europe, Islam still needed the assistance of the medical arts. It developed these to a high degree, as is shown by the fame of its medical schools and its hospitals, and the importance it placed on the knowledge of botany, pharmacy, chemistry and other branches of science, from which much of the modern healing art is derived.

The next four chapters, *Chapters 2* to *5*, tell the story of Rome and the terrible effect of its capitalist civilization on the peasantries and the family, upon the food of the poor, and upon the soil, resulting in its extensive loss through erosion and the formation of marshes. Islam, on the other hand, supported the peasantries, honoured the family, dictated that even the slaves should have the same food as their masters, and took every care to conserve the soil.

Chapter 6 is concerned with the nomads and farmers, and the effect which the scarcity of the food of the nomads had upon the civilizations of the farmers, and the history that resulted.

Chapter 7 contains two contrasting examples; the first being the deprivation of the soil of the Falkland Islands under the dominance of modern commerce, and the second being the renovation of the soil of a Baltic dairy farm through correct farming and ecology. Both are examples, on a small scale, of vital issues. Local self-government in things of the soil would have prevented the first; the second speaks of the final unity of all living things, as in Mohammed's teachings.

Chapter 8 gives an account of the proper and the wrongful uses of urban and rural wastes. Islamic agriculture was, as we have discussed, based on the proper use of wastes.

Chapter 9 continues this theme. It shows why the

dominance of money leads to the wrongful use of wastes, and discusses the difference in thought and values between the rural and urban populations. These differences are illustrated by the different character of their taxations – the natural character of that of the farmers being payment in *kind*, or farm products, while that of the urban centres is payment in *money*. This difference was recognized and acted upon by Islamic civilization. The chapter continues with further illustrations of the economics of the soil, including an account of the Chinese economics of the use of water, as given by Professor King.

It concludes with a summary of the disastrous effect that the dominance of money has upon the fertility and health of the soil.

Chapter 10 contains the story of the peasantry of England and the theft of soil nutrients by the lords of the manor, culminating in the ruin of the peasantry in the Industrial Era. Only the peasants of the Isle of Axholme escaped this fate.

The story illustrates well the lack of decent conduct towards rural labour that came with the introduction of modern capitalism, and is the precise opposite of the sense of sanctity that was bestowed on all labour by Islamic civilization.

Chapter 11 is the first tale of primitive agriculture (that of Kenya), under the aegis of commercial farming, as told by Mrs Huxley. It is a story without redemption; in fact it describes the retribution exacted by the extraordinarily rapid and devastating spread of soil erosion.

Chapter 12 is the second story of a primitive people. The natives in Nyasa were lured or forced from their land to serve in gold and diamond mines. It has an all too similar resemblance to the fate of the English peasantry as described in *Chapter 10*.

The conditions in the Nyasa mines were never as terrible as those in England – where children of six years, harnessed to small carts by chains, drew coal along underground passages – but they suffered nevertheless bondage, drunkenness and disease, under the stress of which many great and successful social and medical improvements have now been effected.

Chapter 13 tells the story of the salvation of Tanganyika, effected by the little tsetse fly, from the rapid erosion that has visited Kenya. It also contains a most promising story of redemption, regarding the proper use of rivers. Instead of forming boundaries of human hostility with the ill-effect that such a river as the Rhine in Europe has had, they are made beneficial, through being used for the benefit of the people on either bank.

The last tale of the effect of the dominance of money over a primitive peasantry is that of our oldest colonies, the West Indies, in *Chapter 17.* The hardships and erosion that the West Indies have suffered is illustrated with an account of the uninterfered with – and flourishing – island of Lombock in the East Indies.

Under the Islamic principles of the treatment of peasants, none of the disasters of *Chapters 11, 12* and *17* could have occurred. On the contrary, examples such as that of Lombock would have been more common, as the results in the Mediterranean Muslim islands show.

Chapter 14 is a philosophical interlude. It deals with the extraordinarily delicate and varied nature of our food substances, and how they are built up from only a few elements, the most common of which are not only earthly, but also aerial. It talks of the need for a wider conception of them and their nature, if positive mental and physical health are to be attained.

Chapter 15 is another tale of the effect of the dominance of money. It concerns itself with Sind and Egypt, and shows the danger to the alluvial soil of trying to force it beyond its capacity.

Perennial irrigation, which has been introduced into Egypt and Sind, has been first financed by money (with interest), and thereby has followed a path different from that of Islam, in which the cost of new canals was borne by the government, and the cost of maintenance and repair alone shared between the State and the users of the water. There was no interest; there were no bankers who brought huge sums of money into existence out of nothing and issued them as loans to be repaid with interest. The first thing the bankers sow upon the new land is debt, as one might sow tares amidst the wheat. In doing so, they enforce greater productive effort upon the land than it can bear.

As Mohammed says:

> 'Although interest brings increase, its end tends to scarcity.'

This saying proves itself as true as ever in this modern example. Both in Egypt and Sind, the land has been forced beyond its natural capacity, and this has already produced scarcity because of the resulting increase in alkalinity.

Writing from a purely agricultural perspective, Professor King says that in all probability the people whom our modern civilization has supplanted knew of this error, through having tried and rejected perennial irrigation. Islam went further in the interests of soil; from the very start, it shut out the men of greed.

Chapter 16 is a second interlude chapter. It is a review of artificial fertilizers. Their introduction was a fragmentation – an incursion of one particular arm of science into the realm of farming. These scientists took only a partial view of the character of the soil. They took a few of the most important chemical elements of the soil, and tried to make them into a whole. They sought to displace natural manure with measured doses of these chemicals; like modern doctors, they first diagnosed the land, and then prescribed for it.

Artificial fertilizers began to be important when the quantity of natural manure began to decrease, due to railways displacing horse traffic, and sanitation abolishing the use of refuse as fertilizer for the soil. With motors displacing horses, and tractors displacing horses and oxen on the farm itself, artificial fertilizers became even more strongly advocated.

They have their occasional place in increasing the productivity of depleted or poor soil, and they have been of great service during the period of war. But they have also distracted attention from natural manures, and so have helped to hide the disastrous effects of the misuse of wastes.

Artificial fertilizers represent a fundamental fault in philosophical thought, one that ignores the wholeness of the relationship between the dead and the living in farming. They have created a farming so beset by disease that the scientific farm has become a blend of factory and hospital, producing products inferior in health, quality and taste, and causing chronic deterioration of the soil.

Artificials, of course, played no part in the farming of the Islamic civilization; nor, indeed, of that of ancient Peru, nor of any of the great farming cultures of the past.

Chapter 18 is the tale of the German colonies, and is one in which the consequences of the belief in the rights of the strongest is consummately expressed. The latter part of the chapter tells the happy stories of these same colonies under the guardianship of the Permanent Mandates Commission of the League of Nations. Differing in manner, three governments – those of the Union of South Africa, France and Britain – effected a miraculous change by means of principles approaching, and in the case of Tanganyika nearly identical with, those of Islam, when dealing with the oppressed peoples of the soil.

In *Chapters 19* and *20*, the countries of Russia, South Africa, Australia and the United States illustrate the climax of the destruction and death of the soil which our modern values make inevitable. In spite of their great scientists, all these countries have been placed in grave danger.

In Russia, erosion, particularly that due to deforestation, has been occurring for longer than in the other three countries of later development. In the case of Russia's modernization, and especially its almost fanatical faith in the tractor and machine farming, Professor Kornev has given the warning:

> 'In the present day, there are huge areas in the USSR where, owing to the excessive breaking up of topography, whole territories formerly under profitable agriculture are now occupied by immense ravines and infertile wastes.'

South Africa has been described by R. Whyte under the heading *The Transformation of South Africa into Semi-desert in the Twentieth Century.*

E. Clayton, in *Overseas Investigation* (1937), states:

> 'There is no doubt that we Australians are in the process of transforming the semi-arid areas into desert at a more rapid rate than in the USA.'

...and in the wetter riverine districts, many parts are gravely affected by erosion.

Finally we arrive at the USA, which has become the leading modern country in the production of food for itself and other countries, as well as of other essential crops. Yet in doing so, it is fulfilling the prophecy of its own Professor Shaler – that unless some radical change is adopted, we must anticipate a time:

> '...when our kind, having wasted its great inheritance, will fade from the earth because of the ruin it has accomplished.'

The chapter closes with an example of a great awakening – a powerful movement towards redemption in the USA; the Tennessee Valley Authority, which follows the values of Islam in the priority given to local, agricultural knowledge, and the balance and mutuality of all labour, whether it be in the factory or on the land.

24

A Plan for Action

THE WORLD, at one time so very large, has shrunk a good deal these days, and a number of questions that were once national or local in character have consequently become global in scope. Certainly, if any question has worldwide significance, it must be that of the treatment of the world's soil.

It follows, therefore, that any action to be taken regarding reconstruction of the soil must also have a worldwide aspect to it.

The present state of the world (*the author is writing during the final stages of WW2 – ed.*) shows us convincingly – and of course we scarcely needed another demonstration – that at a time of war, every nation involved, whatever its form of society and government, is not hindered by money; nor is it hindered by unemployment.

At such times there is work for one and all, and work takes precedence over money; money does not call forth work, but work becomes the master of money. The impulse to action is so great that it takes complete mastery of humanity and matter.

Similarly, to bring about a reconstruction of the soil, a wide-ranging impulse will have to be called into being. It is not to be expected, of course, that it will have the cohesive fury of a people called to war, but it should certainly seek to apply means that are worldwide in their character.

What means of this character, then, are available to a large scale movement such as we need today?

First and foremost there is the unique – and only – positive achievement of the war. The war has achieved, as never before, the technical unity of the world. The spirit of unity is quite a different matter. The riven spirit of the pre-war times remains, and as yet there has been nothing in the planning for the post-war period to prove positively that it has been radically diminished.

This is as one might expect. The last few decades have seen the improvement of technique, to which one great war, one great revolution, expectation of war, and a second great war have jointly given an impulse that has been irresistible.

This technical achievement cannot possibly be overlooked as a means to reconstruction. With its wireless reaching so many homes and papers, its multiple air routes and air bases, its great roads, even amidst the supposedly eternal defiance of the mountains of Central Asia, its innumerable ships, its myriad inventions in means of communication, it has developed a capacity to weld the world together in a way that foretells a new era. The world has been technically transformed into one borderless whole, for neither the air nor the aether has frontiers.

Unless, then, mankind is to be overtaken by physical degeneration, or unless we enter one of those periods of disintegration of civilization with which historians are so familiar, and of which the decline and fall of the Roman Empire has been the classic example in the West – there is no possibility of the future technical disseverance of the world. Its technical unity is a fact, and these days everything depends upon the uses to which it is put.

It may, for example, be made subservient to a revival of the pre-war power of money, through a continuance of the bureaucracy necessitated by the war. It may be limited by fragmentation, if the divided nationalities continue to indulge their political appetites and make the primary needs of their peoples means to national aggrandisement. Indeed, its use and misuse could well lead to further wars, famines and increasing social chaos.

All this could happen if humanity continues to ignore the health of the earth's crust. The uncontrolled spread of the soil's devastation will finally force us all, before it is too late, to use the worldwide capacity of this tremendous technical power for the purpose it seems to be so well suited to – namely, the world's salvation, and not its destruction.

Those, then, who are convinced of the need of reconstruction by the soil should not now allow themselves any laments for the past, or indulge in vain dreams of a world other than it is at this very present.

Those who would reconstruct the world's soils should look upon the technical unity of the world as a gigantic and powerful assistant to the awakening of the brotherhood of humanity through the common parent of all, the soil.

They have both a message, and the means of communication commensurate with its vastness. The message has undoubted power because physically, mentally and morally, it affects all humans who tread the earth.

If the messengers can enforce themselves upon the world's wireless; if they can reach once distant lands by air travel, which now makes east and west and north and south neighbours; if in the world's press they can publish to innumerable readers at one and the same time the news of their movement; if people of each country can communicate to other countries what they are doing, what developments have been accomplished or are expected – *then they will fill the world with the creed of the soil.*

There is much to communicate these days, as soon as the din of war has become silent. When the thoughts of people are no longer directed to the slaughter of life, but to the means of its conservation, there will be many tasks.

Each country will need its own movement of men and women who will take a part in this new unity of the world. The type of individuals required to form the initial bonds is of the greatest significance. It is clear that they must be the type that was kept almost voiceless in the period between the two great wars, when their adversaries, by converting soil fertility so freely into money were, in the words of Jacks,

'drawing the whole world headlong to starvation.'

Upon their ignorant greed, there was, says Jacks, at that time only one check – the threat of war. Then came the reality of war. The governments of the Great Powers, realizing at last the paramount and primary character of soil fertility, allowed it no longer to be turned freely into money, but treated it as a national armament, no less precious than metals and chemicals. This dominance of soil over money must not be allowed to relapse after the end of the war. *The lesson learnt must this time be unforgettable.*

Indeed, a new and non-violent war will break out; a war on behalf of the soil, health and life, and the physical freedom of the people.

In this war, the soil and its allies will have, in the beginning, many opponents.

Firstly, except for men and women of genius, who have the capacity to change their outlook and break with the long-held habits of the past, those who rose to high authority before and during the war may be expected to be opponents, *because of the very fact that they rose to authority under the ruinous values that have brought about the earth's devastation.* To purge themselves of the gross defect of mind that the values entailed will be beyond their capacity; whether they wish it or not, the familiar spirits will not cease to haunt their thoughts and actions.

Then there are aliens, and those without any country

of their own, without either any inborn, native love for the land in which they have their refuge, or any actual kinship of mind, occupation, and tradition with the soil.

With both types of people, money is, of necessity, the paramount object, because the only worldwide rival to money is the soil. Such people are dangerous.

Urban people, in general, are likely to be opponents in the beginning, for the urban population functions as a mass rather than as individuals. In small matters they will hold the opinions of their set or subgroup; in larger matters, they are subject to mass emotion. Their interests and faiths wax and wane, are hot and cold. Fed by news selected for them and spurred by propaganda, they are the objects of unfixed laws, each of which, like a wave on a sandy shore, wipes out the impression of its predecessor.

Separated as they are from the soil and its creative powers by modern town life, urban dwellers are unfamiliar with the ancient notion of fixed and immutable laws, by which alone the dominance of the soil can be maintained.

Even modern education itself is an opponent and a powerful detractor of the land, because the soil is largely regarded by it as something to be ignored. All education for the young in all advanced countries seems to possess this profound defect.

Personally I am best acquainted with my own education, in a public school in England. The school was situated amongst cultivated fields and riverside pastures. Yet never once was the local character of the land held to be of sufficient significance to be mentioned by the teachers.

I realized vaguely, even then, that our education *did not start at the beginning*, not from the soil and the river from which our life began, *but from somewhere else*, as if the roots of being did not matter in education and could be left invisible or unknown to the 'educated' mind.

It taught us to be gentlemen, to regard ourselves as something superior to the soil. It removed us, from the start, from any equivalent to the Islamic notion of the sanctity of the soil. This stigma most of us had to carry throughout our lives; only the rare sceptic escapes from its trammels.

Even when the Empire called some of my colleagues to the charge of 'primitive' agricultural peoples, the same attitudes prevailed. G. Herrington, at a West African agricultural conference (1938), stated:

> 'Unfortunately most of the Europeans who come out to this country (Nigeria) have received an education which is divorced from rural life and few have any knowledge of its interest and variety, or the intelligent skill that rural life entails. This type of education has created an attitude that is very difficult to overcome.'

It has only very rarely been overcome, only in great sceptics or those with rare sympathy for their rural subjects, such as were pre-eminent in British India before the Mutiny: Sir Thomas Munro, Sir John Malcolm, Sir Mountstuart Elphinstone and Sir Charles Metcalfe. Because of this almost invincible attitude, our dominion over rural lands has been one mostly divorced from rural life, and has been overwhelmingly antagonistic to the soil.

In addition to the famous British public schools, education in our state schools, in the United States and other countries, has had the same tendency.

In the German University where I spent eighteen months in post-graduate study, it seemed to me to be the same; men with so-called brains were considered to be suited for something better and more lucrative than work upon the land. This profound fault in the education of the Industrial Era has caused untold damage in health, sanity, food and the conservation of the soil. It makes education undoubtedly an opponent, not an adjuvant to reconstruction by way of the soil.

The new men and women – for the war has brought many women into direct contact with the land – will be those who have been shaped and fashioned by the soil to a form of serenity, a sense of the spaciousness of time, and a capacity for individual judgment.

The soil itself has been their textbook, and printed books are only subsidiary. Books may widen the understanding and give to their students knowledge of many of the chemical and physical properties of the earth's crust, but they do not possess the almost magical power that leads to the revelation of the soil itself. They are very valuable supports and helpers, but they are not initiators of the sense of kinship. Initiation belongs only to the parent of life.

The new men and women know the soil and its creative powers personally, learning chiefly through their eyes and muscle-sense, and not through their ears. Their knowledge and feeling for the soil are the same as they are for other living things – a matter of touch, smell and sight, a physical response to contact with it. It is made up of a variety of factors; the feel and sight of rain, snow, dew, sun and wind; the characters and purposes of hedges, woods, fields, hills, valleys and plains, of insects, plants, flowers, weeds, all subject to the seasons in their progress through the years.

It is, then, something very real, something very vital, something that proclaims an ordered multitude of being, far transcending the ephemeral life of individuals.

Goethe said:

> 'It is to the fresh air of the open field that we belong by right. It is as if the Spirit of God there breathes immediately upon men and thereby a godlike strength exercises its influence.'

The new men and women possess or gain a health that transcends what the practitioners of scientific medicine have taught the public to regard as health – namely, something that can be acquired by a process of discovering and putting into practice separate means of escape or recovery from various diseases.

What will be required is not this piecemeal approach, but positive, whole health, which exists in itself, and is something quite apart from disease. It will be required, because it is a necessary prerequisite of the comprehensive simplification which these times require. 'Cleverness' exists these days in abundance, for when the simplification of positive health is absent, cleverness finds opportunities to express itself in a thousand hydra-headed problems.

It is for this reason that, in spite of the numbers of educated, clever men and women – and in spite of their abilities in dealing with fragmentary social and political difficulties – in their lack of any understanding of what is *really* happening in the world, they have failed entirely to avoid the emergence of a series of catastrophes.

Health, therefore, there must be. The simplification that it brings – just as healthy fields create a healthy humanity – in dealing with a myriad of hydra-headed difficulties is essential.

Health is, as Goethe said of truth, like a diamond; it emits its rays in all directions. Being whole itself, it brings with it a lively valuation of the things of health and wholesomeness, and a ready acceptance of them, along with the rejection of the fragmentary.

Its convictions are not mere matters of mental persuasion. They are matters of bodily response, sober and purposeful in action and hard to oust, for they are creative and positive. By allowing correct choices, they prevent the complication of many particulate and specific solutions burdening a problem with much argument, for they are attracted to reality intuitively. The correct life of the earth is in reality not nearly so difficult or complex as is the wrong, because it is *simple*, in the root meaning of the word; *unity*.

It is to men and women so equipped that the initial guidance of reconstruction should be entrusted. Power comes later, when recognition of the need and urgency has become widespread; then it will be given, as it was willingly and freely given by the mass of people to its leaders during the war.

Of this power, the great urgency of war has provided many valuable precedents, the memory of which will hopefully not die out with the rapidity that the fast pace of modern-day events can engender.

Of these precedents, there are few which surpass that established by Lord Woolton, the Minister of Food in Britain. His is, to my mind, a classic example of an inspiration to workers in countries concerned in these matters. One only wishes that reconstruction via the soil could be almost a continuance of the work done by the Departments of Food and Agriculture, without the long delay that seems inevitable before the public realises the need for reconstruction.

As to the nature of the work that is required, it will come under headings such as the following:

1. The restoration of the peasantries and peasant families as the primary cultivators of the soil; the use of large estates only when suitable to particular soils, forms of cultivation, or social conditions;
2. The freedom of the soil from the power of money;

3. The priority of the soil's claim upon a country's water, and the local control of the distribution of water;
4. A rural education, which is, both locally and generally, a true education in the soil;
5. An education of all urban populations which begins with the soil and the life which it provides to all humanity;
6. The adoption by both town and country of the rule of return;
7. Recognition of the unity of all health – of the soil, plants, animals, and humanity;
8. The right of all people to their share of essential foods and work; and
9. The use of modern techniques in promoting and maintaining humanity throughout the world, through the common bond of the soil and its conservation.

The gates of change have been thrown open by the current war, and when I venture near them to attempt a view of the future, I must confess that I am confronted by a dazzling vision, one which this cruel war seems to be the immediate cause of.

The war has brought together as allies the four powers that have control over the four greatest areas of land on the surface of the globe. These four powers divide themselves by physical proximity into two pairs. First we have China and Russia, with a basic similarity that I will discuss shortly. The other pair, connected physically only by the proximity of Canada, are the USA and the British Empire, the two leading capitalist powers of the world.

These four allies form a strange conjunction, characterised by many differences.

There is first China, with an unequalled history of stability and conservatism, now torn asunder by the inroads of modernity.

Secondly, there is Russia, also an ancient autocracy, which has recently overthrown capitalism and with fierce energy has created the USSR, an authoritarian collective state.

Thirdly, there is the USA, so compact in the spacious unity of its territory. It endeavoured to shut itself away from the troubles of the world, but was suddenly aroused to the futility of this isolation by the catastrophe of Pearl Harbour.

Lastly, there is the British Empire. Its dominions are spread so wide and varied as to make it the leading power of the world, but Britain has been startled at the discovery of its inability to protect its far-flung possessions – and almost its homeland – in the early years of the war.

There are, then, huge differences of need and necessity in the character and circumstances of the four allies.

One general need, however, confronts all four; it is the need to prevent the repetition of the present situation by precluding the possibility, now and forever, of further aggression by the Germans or their Japanese allies.

Apart from this political need it might seem that there is no bond which can inspire genuine unity, and so overcome the manifest differences of character and circumstance between the four allies.

Yet there *is* such a bond – the bond that ultimately unites all humanity in an ultimate similarity, and *that bond is the soil.* It is the soil, and the soil alone, which can bind the four powers together in a reconstruction of life.

All four powers, and with them the rest of the world, are bound together by the perilous condition of the world's soil. None can escape this danger, given the new technical unity of the world. That is the one imperative and vital bond in their conjunction for reconstruction.

Now let us consider the four allies separately, and see what contributions they can make to this fundamental question, and their own particular needs with regard to it.

The Chinese are by far the oldest people of the allies. Their unique contribution is that of the accumulated wisdom of 4,000 years. None have better described this gift of the historic Chinese and their pupils the Koreans and pre-modern Japanese than Professor King, in a partially written *Message of China and Japan to the World*, which he proposed to add to his book *Farmers of Forty Centuries*, but was prevented from doing so by his death.

In the part of this message that has survived, he wrote:

> 'It could not be other than a matter of the highest industrial, educational and social importance to any nation that it should be furnished with a full and accurate account of all those conditions which have made it possible for such dense populations to be maintained upon the products of the Chinese, Korean and Japanese soils.
>
> Many of the steps, phases and practices through which this evolution has passed are irrecoverably buried in the past, but such remarkable maintenance attained centuries ago and projected into the present with little apparent decadence merits the most profound study.
>
> Living as we do in the morning of a century of transition from isolated to cosmopolitan national life, when profound readjustments, industrial, educative and social, must result, such an investigation cannot be made too soon.'

The practices and the methods by which these meticulously careful farmers conserved the fertility of their soils are nowhere better described than in King's book. However, he makes no specific mention of the Tsing Tien system of which Dr Ping-Hua Lee, in Volume 99 of the Columbia University's *Studies in History*, wrote:

> 'The whole history of the government administration of agriculture in China coincides with the Tsing Tien system. Its vicissitudes, its crises and its epochs were timed by the abolition or re-establishment of the system.
>
> It is fortunate for the economic historian that the Tsing Tien system is coincident with China's political history.'

Yet, not the Chinese farmers' devotion to the rule of return; not their incomparable and tireless spreading of the mud of their numerous canals to the extent of 70 tons per acre; not their careful preservation of the humble earthworm, which, said Darwin, spreads ten tons per acre of an even finer soil than silt upon the fields, in addition to the other services it performs; not the irrigation of their carefully levelled fields; not the mixed crops – none of these will form a bond more firmly riveting peasants to peasants, and peasants to the land, than this Tsing Tien System.

The *Kolkhoz* (collective farm) system of the USSR is a varient of the Tsing Tien system, modified to address the need to provide food and other agricultural products for the many new manufacturing towns which have been built to secure the USSR's place amongst the modern powers, and to equip them for the war, for which their rulers have prepared with such great speed. The Russian farming families have the same private fields handed over to them for continuous ownership and their own subsistence. This central plot, larger though it is than the ninth field of the Chinese sages, is similar to that ninth field, in that it is the State's plot, worked co-operatively by the farming families.

Moreover, the very dangers of the *Kolkhoz* system, in the pressure that has been put upon it for large and speedy returns by the threat of war, will find their solution nowhere better than in a study of Chinese methods. There is potentially no stronger bond between two huge, neighbouring populations, the Chinese and the Russians, than this bond of their peasantries.

The Chinese themselves are also in great need of effective bonds with their allies. The Chinese leaders eagerly look to the engineers of the British Empire and the USA for instruction in how to deal with such problems as the devastating floods of their great rivers, especially the Yellow River; to reforest their barren catchment areas; to repair the fertile loess soil which was the teeming home of their first ancestors, but is now so miserably given over to waste; and in a hundred other ways to assist in the reconstruction of a distraught farming people.

None of the allies, then, has so much to give – and so much to receive – in the matter of the soil as the Chinese.

Among the four Powers, the Russians are the next oldest people, after the Chinese. Ivan IV (1533–84 A.D.), also known as Ivan the Terrible, is now heralded as one of the most important leaders of the Russian people. With ruthless determination, he consolidated the Russian lands, drove out the Mongols, made the Volga into a Russian river, annexed Siberia and made its lands so attractive to the Russian peasants that his successor, Boris Godounov, had to issue an order stopping further migrations of peasant families.

In *Chapter 19* we saw how the Russians had eroded vast areas of their land from the time of Ivan onwards, mainly through the destruction of forests in order to open up new land. They did not even spare the watersheds and their slopes. The tempo of those days was far slower than that of the modern Russians, whose need for cash for foreign machinery has led to the wholesale destruction of Russian and Siberian forests.

The Steppe and other lands did not offer the same inducement, so, while they were developed and arid lands were reclaimed with singular skill, it was largely the forested lands which were gravely depleted. To the warning of Professor Kornev, quoted by Jacks and Whyte, the two authors added this comment:

> 'The tractor plough is the enemy of grassland in dry areas, but is indispensable to the propagandist in the modernization of Russian agriculture. Though fore-warned by the experience of other countries, it is difficult to ascertain whether the authorities are aware of the dangers of mechanization.'

To what degree the Russians have degraded their farmed soil because of the pressure of the war cannot yet be known, but it must be considerable; it may be, indeed, the greatest loss which they have suffered, despite being among the victors in the war.

That they have much to give and much to receive from their allies in terms of the soil is clear. They can share their experiences, particularly in the reclamation of arid land, and they can give a picture of land developed under an economic system by which the land is developed and farmed without the burden of financial debt, but with the help of the State's revenues. They were about to institute, in their fourth Five Year Plan, a wide-ranging control of water, linking together the waters of the north-flowing and south-flowing rivers, as well as individual rivers, and the same with the rivers flowing east and west. This is, perhaps, to use Professor Kornev's words, an act of the excessive breaking-up of the topography, the results of which can only be estimated by experience. Yet, on a small scale, Mesopotamia once provided an example of unparalleled success resulting from the linking of rivers.

So the Russians have much to give in terms of the soil. They have also much to receive; especially from the great work of the Americans in the reclamation and conservation of arid lands, and the regrowth of forests upon watersheds. They have, then, many bonds to forge in terms of the soil.

The last two allies are powerful examples of the development of farming under the dominion of money. The picture that they provide has already been sufficiently illustrated in previous chapters as one of progressive destruction of life for the sake of temporary benefit to financial farming and ranching. The Americans and British have both pursued the path of ancient Rome with a tempo far exceeding that of the Romans. One writer, indeed, has stated that North America would, at the present rate, be turned into a Sahara within a century.

The fact is that neither the Americans nor the British are yet civilized in terms of the soil; neither has learnt anything of the wisdom of the East. Their use of money is too often nomadic. They invest it in land or other ventures for personal profit and, when profit fails to appear, the pressed and overworked land is abandoned, and the money transfers itself to other ventures, even such as granting assistance to other countries to arm themselves in preparation to fight the very countries and peoples of the lending financiers themselves.

Between the financiers and the nomads, there is, indeed, very little difference in principle and in values in terms of the soil. Granted, the nomads of the past risked more; they risked their lives, and those of their wives and children, in fact their very existence.

The modern financial nomads, on the other hand, are

consciously risking very little. They have risked much *unconsciously*, though; their own lives by enemy bombing, as well as those of their families, plus their homes, and the very safety and freedom of the countries in which they live.

Many farmers in the United States have recently become woefully aware of, and alarmed at, the effects of this nomadism on the soil – of the unrestrained clearing of forests, the overgrazing of the deforested land, of the deep ploughing and mechanical farming of their prairies, of the one-year tenures of farms which enable men to turn fertility into cash and, when this land is degraded, to purchase new land in the great territories of fertile soil which remain.

They can ponder over special maps, as we did in *Chapter* 6 over the map of Asia, and read their own fate there. Such a special map of the USA is to be found on page 51 of Jacks and Whyte's now famous warning to the world. The land of little or no erosion is white, and the eroded lands are graded in shades to indicate its character and degree. One may well shudder at the supreme peril of this great people, if one considers the implications for humanity of this map. The white areas are so few; they seem to cover but a tenth of the map. The rest is eroded lands, according to their kind and their degree, together with mountains, mesas, canyons, and badlands. This visual evidence is enhanced by a number of photographs which are enough to terrify the mind, eased though it may be by the knowledge that what is happening points to a dreadful future to which our own lifespans will not extend.

Against this tale of destruction, with the haunting fear it brings to farmers of the richest country in the world, can now be set the supreme achievement of the Tennessee Valley Authority, which has brought about in the valley's inhabitants a veritable resurrection of the human spirit.

Lilienthal does not fail to recognise this great change,

and quotes the editorial of a newspaper from Alabama, one of the seven states covered by the TVA:

> 'We can write of the great dam... of the building of home-grown industry and of electricity at last coming to the farms of thousands of farm people in the Valley. Yet the most significant advance has been made in the thinking of a people. They are no longer afraid. They have caught the vision of their own powers.'

This gift of confidence, one thinks, will be the outstanding contribution of the USA. It has already attracted the keen interest of many governments of South America, Europe, the East, and South Africa.

But if the USA has much to give, it also has much to learn from its allies. From the Russians can be learned the value of saving the soil from the dominance of money; from the Chinese, the meticulous conservation of the soil, the full rule of return, and agriculture as a national art; and since they now have colonies; from the British, the right and wrong ways of the treatment of tropical colonies, and much else.

The gift of the British to their allies is, indeed, unique in that they, with their extensive empire, have been brought into contact with all kinds and varieties of agricutural conditions. The British have a vast knowledge of the world; ranging from the many millions of India and their imperilment due to their relation of the soil, which I have described elsewhere in my *Restoration of the Peasantries* (1939); from the vast plains of Canada, which share so fully the most dramatic dangers of the Dust Bowl of the United States; from the disasters that are afflicting her Australasian colonies; from the erosion and degradation of the fertile West Indian Islands, to the twin islands of the Falklands in the cold waters of the South Atlantic.

They are indeed fortunate, in comparison to their allies, in that their homeland, set in a temperate sea and served by a humid and largely mild climate, is almost free from the dangers of erosion. The problems of the reconstruction of Britain's misused soil do not have the gigantic proportions of the homelands of the other three.

The British, therefore, with the wide variety of their experience, can act as a bond not only between the Allies, but between them and the wider world. Their knowledge and understanding of cold, temperate and tropical soils and their peoples is not yet deep enough for them to be absolutely the source of unity that is needed, but at least they possess the links through which that knowledge, when recognized and formulated, can be diffused. With their contact with their allies and their own colonies and their peoples, with their empire-made neighbours, the British, like their great technical achievements, can have a profound influence on the world.

The coming meeting of the four Powers, then, has a potentially far greater meaning than a 'Conference of Powers' might otherwise usually have. It is a meeting, not just of tongues and diplomats of the countries they represent, but of the soil of the world and of mankind. It is a meeting, not of four great powers only, but of four great masses of humanity, all witnessing the rebellion of the soil against its treatment by humanity.

The participants do not include just China, but the Chinese with their 40 centuries of farming; not just Russia, but the Russians, who first united great tracts of two continents into one whole and who are now testing ways of treating their soils so as to form the basis of a civilization of stability; not the United States, but Americans, who are gathering their forces together to stem the terror of an insulted virgin soil; and not just Britain, but the British, who have been marching upon the path of soil destruction so clearly marked out by Rome, but who, with a courage and enterprise that can only belong to a great people, link together most parts of the habitable world.

What a conjunction of opportunity!

The heart almost stops at the thought that, had the war ended as at one time it seemed it might end, the future of the world would have belonged to the Germans, who, shut in their history between their southern and eastern neighbours and the cold northern seas, have none of the vast treasures of experience which the four Allies can bring to reconstruction.

To the four Allies are opened the gateways of an opportunity to bless the whole world as never before. Beyond the murk and rubble of this terrible war; beyond the last, bleak resting places of millions of heroic men and women; beyond the razed homes and shattered towns of 'the quiet people'; beyond the scorched acres and barren fields; beyond the famines and their reign of death; beyond all this horrible orgy of destruction, appears the vista of the living earth as the source of the reconstruction of mankind.

At the gateway to this reconstructed future stands a sentinel, awaiting the only password it will accept:

THE SOIL

ALSO FROM A DISTANT MIRROR

The Blood and its Third Element
Antoine Bechamp

Bechamp or Pasteur?
Ethel Hume

Reconstruction by Way of the Soil
Guy Wrench

The Soil and Health
Albert Howard

The Wheel of Health
Guy Wrench

The Soul of the Ape & My Friends the Baboons
Eugene Marais

The Soul of the White Ant
Eugene Marais

Earthworm
George Oliver

My Inventions
Nikola Tesla

The Problem of Increasing Human Energy
Nikola Tesla

Response in the Living and Non-living
Jagadish Bose

WWW.ADISTANTMIRROR.COM.AU

www.ingramcontent.com/pod-product-compliance
Lightning Source LLC
Chambersburg PA
CBHW020913020526
44107CB00075B/1669